HANDICRAFT PHILOSOPHIES

Handicraft Philosophies

CRAFT, REPRESENTATION,
AND SOCIAL KNOWLEDGE
IN EIGHTEENTH-CENTURY BRITAIN

Ruth Mack

STANFORD UNIVERSITY PRESS
Stanford, California

Stanford University Press
Stanford, California

©2025 by the Board of Trustees of the Leland Stanford Junior University. All rights reserved.

This book has been partially underwritten by the Peter Stansky Publication Fund in British Studies. For more information on the fund, please see www.sup.org/stanskyfund.

No part of this book may be reproduced or transmitted in any form or by any means, electronic or mechanical, including photocopying and recording, or in any information storage or retrieval system, without the prior written permission of Stanford University Press.

Library of Congress Cataloging-in-Publication Data

Names: Mack, Ruth, author.
Title: Handicraft philosophies : craft, representation, and social knowledge in eighteenth-century Britain / Ruth Mack.
Description: Stanford, California : Stanford University Press, 2025. | Includes bibliographical references and index.
Identifiers: LCCN 2024042821 (print) | LCCN 2024042822 (ebook) | ISBN 9781503606371 (cloth) | ISBN 9781503642935 (paperback) | ISBN 9781503642942 (ebook)
Subjects: LCSH: Literature and society—Great Britain—History—18th century. | Knowledge, Theory of—Great Britain—History—18th century. | Enlightenment—Great Britain. | LCGFT: Literary criticism.
Classification: LCC PR448.S64 M33 2025 (print) | LCC PR448.S64 (ebook) | DDC 121.0941/09032—dc23/eng/20241031
LC record available at https://lccn.loc.gov/2024042821
LC ebook record available at https://lccn.loc.gov/2024042822

Cover design: Gabriele Wilson
Cover art: Primary piece: © Mary Evans Picture Library. Border art: William Hogarth, *The Analysis of Beauty, Plate 1* (detail), 1753. Etching and engraving; third state of three. 15 1/4 x 19 11/16 in. Harris Brisbane Dick Fund, 1932, The Metropolitan Museum of Art, NYC.

The authorized representative in the EU for product safety and compliance is: Mare Nostrum Group B.V. | Mauritskade 21D | 1091 GC Amsterdam | The Netherlands | Email address: gpsr@mare-nostrum.co.uk | KVK chamber of commerce number: 96249943

Contents

	Acknowledgments	vii
INTRODUCTION	Making Social Knowledge	1
ONE	Education *Crime, Punishment, and Realism in Defoe*	31
TWO	Labor *Craft Poetry in Duck and Collier*	63
THREE	Art *Handicraft Lines in Hogarth*	95
FOUR	Science *Practicing Science in Banks's Tahiti*	128
FIVE	Use *Useless Bodies in Johnson and Boswell*	158
SIX	Fetish *Equiano's Antislavery Craft*	194
CODA	Handicraft Humanities?	230
	Notes	233
	Index	277

Acknowledgments

This book is the result of years of conversations, question-and-answer sessions, and work exchanges. I'm grateful for all of it. Audiences at Birkbeck, Yale, Harvard's Mahindra Center, the University of Chicago, Northeastern University, Brandeis, Rutgers, and Berkeley helped get parts of this project off the ground. I'm especially thankful for the workshop gatherings at the University of Western Ontario, Rutgers, and the University of Freiburg (and their organizers Katharina Boehm, Mary Helen McMurran, and Alison Conway), ideal settings for shared thinking. My year at the Radcliffe Institute for Advanced Study was influential for the book, and I want to thank Lucia Allais, Caroline Jones, Ewa Lajer-Burcharth, and Sophia Roosth, who helped me grapple with ideas at a pivotal moment.

I'm lucky to have such wonderful interlocutors within the world of eighteenth-century studies (and beyond). Conversations with Tita Chico, Peter DeGabriele, Helen Deutsch, Sarah Ellenzweig, Sophie Gee, Noah Heringman, Sarah Tindal Kareem, Heather Keenleyside, Jonathan Kramnick, Crystal Lake, Sandra Macpherson, Wolfram Schmidgen, Sean Silver, Courtney Weiss Smith, and Eugenia Zuroski all made their way, in some form, into the book. So too did discussions from graduate seminars, whose members made a mark on this project as its first ideas began to emerge. Here in Buffalo Dave Alff consistently helps me see the field in new ways. As this book came to a close, Elizabeth Mazzolini provided camaraderie and a tight schedule when I needed them most.

I'm deeply grateful for Lynn Festa's rigor and clear-sightedness in her reading of the manuscript for the Press. Thanks too to the anonymous reader for the Press and to my editor, Caroline McKusick, who picked up this project late in the game and saw it through with me. Jude Fogarty and Kevin Joel Berland, along with the team at Stanford (especially Andrew Frisardi and Emily E. Smith), helped make the manuscript into the book you see in front of you.

In the last stages of this project, I counted on some of you who've been there from the start. Theo Davis cheered me up when things went on too long; Janet Sorensen helped me to the big picture and to Rome; and Helen Thompson made last-minute revision queries into pleasurable intellectual work. You are bright lights.

Research for *Handicraft Philosophies* was supported by fellowships from the Radcliffe Institute for Advanced Study, the University at Buffalo Humanities Institute, and the UB Institute for Research on the Study of Women and Gender. An early version of part of Chapter 5 was published as "The Limits of the Senses in Johnson's Scotland," in *The Eighteenth Century* 54.2 (2013) 279–94. Copyright © 2013 University of Pennsylvania Press. Reprinted with permission of the University of Pennsylvania Press.

Several communities within Buffalo have helped me consider and reconsider my own relations to forms of practical knowledge. Among these, I'm lucky to have had the people attached to PS 32, Bennett Park Montessori Center, where I've learned about community, advocacy, and handwork. My students and friends in the Buffalo activist community give me hope for what comes next.

In the many ups and downs of this project, I have had the terribly good fortune to work with Rachel Ablow's advice, hard questions, and unwavering support and amid Simon's and Esther's joy.

HANDICRAFT PHILOSOPHIES

INTRODUCTION

Making Social Knowledge

MR. SPECTATOR, THE FICTIONAL CHARACTER created by Joseph Addison for his periodical, *The Spectator*, shows us the challenge of knowing society. Addison and Richard Steele's *Spectator* papers were among the most popular writings on London society in the eighteenth century, covering everything from politics to wigs and party patches, to the new credit economy. If we wonder what kind of spectator can bring all this to the public, the papers let us know that too. Mr. Spectator, we find out, is tied to a "small Hereditary Estate"; he attended university and can read classical and modern texts; and he has taken a Grand Tour of Europe.[1] His gentlemanly credentials are intact. But if this all seems perfectly ordinary, Mr. Spectator's behavior along the way is not: He reads but does not speak, telling us that he does "not remember that [he] ever spoke three Sentences together in [his] whole life" (*S* 1:2). In London, now spectating for his essays, he makes an "Appearance" but merely listens (*S* 1:3). In fact, he prides himself on not doing much of anything but looking and thinking, adopting the identity of the "Speculative Statesman, Soldier, Merchant and Artizan, without ever medling with any Practical Part in Life" (*S* 1:2).[2]

Addison parodies a new model of knowledge: the scientist's "modesty," his demonstration of truth through control of his body, through the remove of gentlemanly distance and disinterest.[3] But if this way of truth making might be at home in the laboratory, it proves a quite unworkable epistemology for the social world. To perform his duties, Mr. Spectator doesn't just become a "Looker-on"; he has to lose his material, human body altogether, becoming a kind of ghost (*S* 1:5).[4] In Addison's configuration, this is the result of an epistemological mistake. Mr. Spectator tries to have two things

1

at once: the bodily remove that justifies knowledge and the bodily presence and activity that marks him as a fellow member of society. The problem, as Addison sets it up, is one of theory and practice. Mr. Spectator is a "Speculative Artizan," a theoretical handicrafter, defined by a contradiction in an epistemology that claims to grab practical, immediate, hands-on knowledge but without getting close.

Addison opts not to resolve Mr. Spectator's observational troubles in the origin story of his periodical, but the thinkers and makers of *Handicraft Philosophies* pick up this challenge. The writers I study in this book take as central the big question behind Mr. Spectator's absurdities. What kind of "observation" can bring society into view? A century or more before sociology and anthropology offer disciplinary answers to this question, writers begin where Mr. Spectator does: with the terms that natural philosophy offers for an experience of the natural world. But they are not "lookers-on." The "social knowledge" of my title stands for a group of answers eighteenth-century writers give to this new problem of knowing society by doing what Mr. Spectator refuses to do: using the body to know.

My title, *Handicraft Philosophies*, forces together two forms of knowledge that, centuries later, still seem radically distinct. Philosophy, after all, is an abstract, culturally elevated way to know, the stuff of books and an education of the individual mind. Handicraft, by contrast, pulls us toward the opposite pole, working through the material world, productive, associated with imitation and apprenticeship, a practical education of the body that can happen collectively. In trying to make use of both kinds of knowledge, Mr. Spectator's "speculative artisan" calls up an old philosophical problem: the relation between theory and practice. When Francis Bacon founds the new natural philosophy (the "new science") by privileging the artisan and "experience," he seems to resolve this classical division and put practice first.[5] The terms of my title reflect this aim, as well as the many philosophies of "handicraft" that follow. "Handicraft" is a common English translation of Aristotle's term for practical knowledge as productive making, *techne*. Thomas Sprat uses "handicraft" to show the Royal Society's Baconian commitment to the artisan in the *History of the Royal Society*, where it sutures philosophical to practical knowledge. (In the book I often use *craft* as a synonym.) Such knowledge passed from hand to hand, body to body, through imitation and generations of workers in the workshop, which is also the site of oral knowledge and collective work.

But Bacon's (and Sprat's) adoption of craft knowledge for the philosopher has some built-in challenges, and Addison exploits one of them, calling our attention to the strangeness of an artisan-based philosophy whose knowledge depends on gentlemanly modesty and remove. In Addison's account, the natural philosopher is a "speculative artisan" who does without doing, who makes without touching the material world. Addison also indicates that Mr. Spectator's quandary is both a philosophical problem (how to combine two different kinds of knowledge) and a social one (who gets to do the combining). After all, Mr. Spectator ends up at this impasse because his gentlemanly body requires a remove that natural philosophical standards seem to oppose. This, again, is a new incarnation of an old problem: from Aristotle onward, embodied knowledge is tied to particular social positions: not only the artisan but the woman, the child, the enslaved person. These old social associations would prove hard for artisanal knowledge to shake.

In the century that followed the foundation of the Royal Society (1660), this unresolved theory/practice binary and the (continued) cultural denigration of embodied knowledge made it impossible for the Royal Society to democratize as it claimed it could.[6] Despite its declared intentions, it remained a society of gentlemen, not practitioners. Yet what seems a problem for the institution, maybe even for natural philosophy writ large, also provides an opening, a resource for those who embrace bodily ways of knowing. The most elite sphere of philosophizing has a commitment to practice, an aim to suture science to craft. Even if one did not join the Royal Society, this last goal was fair game. The authors in *Handicraft Philosophies* write from the position of the artisan—the thresher (Duck), the navigator (Equiano), the engraver (Hogarth). In so doing, they expand the positions from which "philosophy" can be written. They also use their various social positions to interrogate the privileged body on which "experience" rests. The writers in this book include gentlemen (Banks, Boswell) and an artist and engraver (Hogarth), but also a washerwoman (Collier) and a formerly enslaved man (Equiano). Across these positions, practical, embodied knowledge does not remain neutral. Practical knowledge theorized from the position of the formerly enslaved man or the laboring woman takes for granted that "experience" is not universal but determined by society's requirements for entry; a turn to the body provides an opportunity to specify the implications—the problems, the possibilities—of that fact.

The "social knowledge" of my title, then, turns two ways in the pages that follow. First, with Mr. Spectator, writers consistently find society an odd object to know. The conventional narrative within the history of ideas tells us that "science" is turned on "society" in the Enlightenment, in what is sometimes referred to, following David Hume, as the new "science of man."[7] We see this attempt to project the certainty of mathematics or natural history on society, as in Thomas Hobbes's *Leviathan* (1651), where geometry drives the skill of "making and maintaining Commonwealths"; in Montesquieu's *Spirit of Laws* (1748), where governments are taxonomized; or in the Scottish conjectural historians' use of principles of causation that go back to Newton.[8] The "social sciences," as they are called now, are usually understood to have been formulated in this period.[9]

The writers in this study, by contrast, are less sure that philosophy formulated in relation to the natural world can work as well in relation to the social one. This is particularly true when one moves from studying the structures of societies and tries to report on an experience of them, through "customs and manners," a catch-all term for the elements that make up what Hume refers to as "common life."[10] This common life is full of embodied practices, whether we are looking (as Mr. Spectator is) at English society or, as Joseph Banks will, at Tahitian society. These practices—what people do and how they do it—provide a great challenge for the observer, then and now. What kind of thing are they? And how are they to be known? Can one attempt stand outside them and view them politely (Mr. Spectator's question), or can one only know them up close, through practice and participation? The chapters of this book record a set of efforts that find the observation of natural philosophy, as well as its definition of knowledge, unsatisfactory as an answer. Craft, the sensual, experience-based knowledge at the origin of that natural philosophy, by contrast, is well accustomed to grasping bodily experience, even that acquired outside an individual's consciousness.

Second, as I have already suggested, this "social knowledge" has another valence, in the bodily knowledge of a body made by society. Philosophers of practical knowledge, embracing the body rather than rejecting it, sometimes entertain the possibility that society has already made the body that knows. This doubleness serves not as a disqualification for knowing, but as an occasion to consider the dimensions of a socially defined personhood, one that (like craft knowledge itself) might exceed the individual person.

The writing (and drawing) of handicraft philosophies occurs in an eighteenth-century British context with highly visible makers and a print market full of maker's knowledge. While at our own historical moment, "handicraft" and "craft" are thoroughly associated with a precapitalist nostalgia, craft was a present and visible part of the emerging capitalist society of eighteenth-century Britain. The period was saturated with what we would now call "applied science," and what were then known as the "useful arts." In print, eighteenth-century how-to guides were published on topics ranging from perfecting grammar and writing poetry (e.g., John Smith, *The Printer's Grammar*; Edward Bysshe, *The Art of English Poetry*), to making saltpeter (a component of gunpowder) (Israel Aber, *The Art of Manufacturing Saltpetre*), and tending bees (John Keys, *The Practical Bee-Master*).[11] Such texts are dedicated to what Gilbert Ryle calls "knowing how," a knowledge gained by doing (and opposed to "knowing that" something is the case).[12] Their titles frequently refer to "practice" and to "art," the latter in the sense that Samuel [Johnson] defines in 1755 as "not taught by nature," "science," and "trade." Johnson's examples of "art" range from "dance" to "the *art* of making sugar" (citing the chemist, Robert Boyle).[13] When Edward Bysshe presents his *Art of English Poetry*, then, he means art as a contrived "doing," not as the fine "art" later readers might at first assume.

Enlightenment Britain ran with the assumption behind the Royal Society's projected "History of Trades," that natural philosophical knowledge could be gleaned from the artisan, systematized, and used to advance commercial progress and trade. In fact, although histories of Britain in this period, following Neil McKentrick, John Brewer, and J. H. Plumb's influential argument, have tended to focus on its "consumer revolution," the period also sees a kind of maker's revolution.[14] In the eighteenth-century context of how-to, artisans commonly authored their own texts, rationalizing workshop practice. Such knowledge both fueled Britain's imperial aims and was the product of them: as Rajani Sudan has argued, and as we will find in Joseph Banks's example, British natural philosophers took knowledge from other cultures and claimed it for their own.[15] The resulting "knowledge economy," economic historian Joel Mokyr tells us, was vast.[16] The artisan was visible and current, part of the new capitalist world in its tradesmen who were still makers (or closely tied to them) and in the content of the knowledge base for technological improvement, the printed and publicly available texts of how-to knowledge that once were called "books of secrets."[17] As Mokyr has argued, the public availability of the artisan's knowledge laid the foundation for the Industrial Revolution.[18]

All of this should change our associations with "Enlightenment," a term that still carries its tie to the grand philosophical tradition that in Britain moves through Bacon, John Locke, and David Hume. Locke himself, near the end of his treatise on education, recommends that the gentleman take his free time to become a *"Carpenter, Joyner,* or *Turner."*[19] The British Enlightenment was full of practical knowledge, often well at home in spheres beyond its lowly origins. Ephraim Chambers, author of the *Cyclopaedia* (1728), the nation's first great knowledge project, was trained as a globe maker. Samuel Johnson, known for his *Dictionary* (1755) and his poetry, was acquainted enough with manufacture that he could lecture extensively on coining and brewing. In his educational tracts, Daniel Defoe, author of the early novel *Robinson Crusoe* (1719), shows off his knowledge of cheese and brick making. The *Georgics* of Virgil (29 b.c.e.), the poetry of the greatest classical "speculative artisan," saw renewed attention in this period and was adopted as practical advice for farming, while newer poems in the mode instructed readers on making cider and shearing sheep.[20] A thresher from Wiltshire was considered for the position of poet laureate, in part based on his own georgic poem. From practice to theory and back again, Virgil's poem—which had just seen a blockbuster translation by John Dryden—emblematizes how close the knowledge of the head could seem to the knowledge of the hand, poetry to tilling a field.

The same fluidity is indicated by the table of knowledge from Chambers's *Cyclopaedia*. Chambers has only two major divisions of knowledge: "*Natural* and *Scientifical*," on the one hand, and "*Artificial* and *Technical*" on the other. From our later standpoint the natural grouping of meteorology with algebra might seem familiar, though the presence of law and politics in this "natural" register will not. Most interesting for the purpose of this study is what counts as "*Artificial* and *Technical*," which Chambers is clear involves an "Application of Natural Notices to further Purposes." On this side we find architecture, fortification, navigation, agriculture, hunting, grammar, and poetry.[21] "Applied" or "technical" knowledge, broadly conceived, was frequently referred to in this and in earlier centuries as "art."

The theory/practice stakes of "Handicraft philosophy" become especially vital in the changing knowledge world that follows Chambers.

British society transforms rapidly in this period, with the emergence of a market economy and a middle class, a drive toward trade and empire, a rise in literacy, and an increasingly secularized public sphere. Such changes begin to divide knowledge into parcels more familiar from a twenty-first-century standpoint. To take one of this book's central examples, the new middle-class public and decline of patronage fueled an art market and the establishment of new cultural institutions like societies and museums. When Joshua Reynolds and others criticized William Hogarth, the painter and engraver, for being too much a craftsman (and too dedicated to commerce) for the world of the new Royal Academy of Arts, they pushed back on knowledge classifications that had held for centuries and had made "engraving" and "painting" part of the same knowledge world.[22] Larry Shiner dates the beginning of the breakdown of "the old liberal arts scheme" to 1700 and posits 1750 as the date by which "fine art" began to be consolidated, as institutional mechanisms for this consolidation appeared throughout Europe (including Britain's Royal Academy).[23] When Hogarth and Stephen Duck (the "thresher poet") insist on a broader definition of "art," then, they attempt to navigate the shifting boundaries of knowledge at the moment of this split. They also remind their readers that treating craft as embodied knowledge opens new ways to consider the modern, social world.

As a literary historian confronting this topic, I find my object of study in the forceful moves that philosophers and artisans alike make in attaching their projects to representation. The new experimental philosophy insists that its theory-practice ideal comes with a particular mode of representation: plain language. And artisanal writers of all kinds grapple with the Enlightenment pressures to write the knowledge of the body. In what follows I show how questions around representation are central to natural philosophy's theory-practice claims from the outset and how philosophies are spelled out in the space of representation, a space that can change from one genre of writing to another, each with its own rules for how experience should be specified.[24] I am curious about what happens when an artist, or a botanist, or a navigator, turns to the written word, and I seek to gauge how writing works as both a limitation and an asset for writers attempting to grasp experience that is notoriously resistant to reason, language, and even the mind itself.

Handicraft Philosophy

In the history of philosophy "handicraft philosophy" begins as an oxymoron that becomes a methodological imperative. The British Enlightenment inherits its understandings of theory's relation to practice from both of these contexts: the classical establishment of what would become a long-standing cultural divide between theory and practice; and Bacon's claim to resolve that binary in his experimental natural philosophy. Let me start with the long-standing context in which "handicraft" and "philosophy" are at odds, most famously in Aristotle, whose philosophical program dominated study of the natural world until at least the seventeenth century.[25] Aristotle separates theoretical knowledge from practical knowledge. Theoretical knowledge (episteme) is defined by certainty; evidenced most clearly in geometry, it requires a deductively organized body of knowledge. Craft or skill is a kind of practical knowledge (techne) that is defined by contingency and particularity. Techne is broad in Aristotle's account of "expertise," encompassing everything from making shoes to painting pictures. As the example of the shoemaker suggests, techne is also defined by its connection to productive bodily labor.[26] Aristotle's binary, in which theory is set over and against practice, also maps onto social divisions. Philosophy and theory are the stuff of the mind and of male citizens; practice or techne is the stuff of the body and of women, children, and enslaved persons.[27] The original distinctions are ancient, but we see the same kind of separation well at home in our more modern world, in the divisions that are made and sustained between manual labor and intellectual labor, between work with the hands and work with the head.[28]

Bacon, and, after him, the Royal Society scientists, founded experimental philosophy on undoing this division between theory and practice, overturning Aristotle to position techne at the core of the new philosophy and its way of knowing the natural world. Indeed, for Bacon, artisanal knowledge is scientific knowledge; an examination of natural forces under contrived conditions, craft practice is experiment *avant la lettre*. In the *Advancement of Learning* (1605), Bacon contrasts the scholastic philosophers "shut up in the cells of a few authors," spinning "those laborious webs of learning which are extant in their books" with the natural philosopher whose "wit and mind . . . work upon matter."[29] There is "labor" in both cases, Bacon shows us, but only the experimentalist's amounts to techne, transformative and productive labor. Bacon, calling up the great artisan Vulcan, indicates that there may be two kinds of natural philosopher: he who searches for causes and he who is interested in the "production of effects," the digger and he who "can refine and hammer."[30] Bacon understands philosophy as "maker's knowledge," the knowledge-in-doing of Vulcan or Hephaestus or of the more modern craftsman.[31] Natural philosophy, he says, should follow the "mechanical arts," which "grow and improve every day as if they breathed some vital breeze."[32]

Bacon's engagement with the artisan is not just conceptual, an attempt to specify key differences between his new experimental philosophy and scholasticism. Bacon also sees artisanal knowledge as containing knowledge about the natural world, which the natural philosopher has not yet gathered. In his account of useful arts in his projected "histories of trades" project, Bacon proposes to study artisans in their workshops and to compile the knowledge from those "preferable arts . . . which present, alter and prepare natural bodies and the materials of things."[33] By these "arts" Bacon has in mind "agriculture, cookery, chemistry, dyeing, the manufacture of glass, enamel, sugar, gunpowder, fireworks, paper, and suchlike."[34] Bacon's "histories" will thus advance not only scientific knowledge but also commerce. Although Bacon's project was never brought to completion in its original terms, a number of prominent early Royal Society Fellows—including John Evelyn and Robert Hooke—offered proposals for a "History of Trades." As Celina Fox tells us, "Six histories were begun in 1660, a further eleven in 1661 and five more in 1662."[35] After this time, the project peters out, and the *Philosophical Transactions*, published by the Royal Society, shows mechanical knowledge taking a backseat to medicine and natural history.[36]

If artisanal knowledge had started to slip from the Society's immediate research focus by late century, the artisan and his work remain critical for accounts of the *kind* of knowledge the Society produces. Thomas Sprat uses Bacon's terms to stress the Society's commitment to craft knowledge in his *History of the Royal Society* of 1667. Sprat had joined the Society in 1663 and had almost immediately been tapped to write the *History*, likely in collaboration with John Wilkins, to explain the project of the Society to the public.[37] Sprat relies heavily on the artisan in his description of experimental science, as in this account of "perfection": "the *Mechanic Laborers* shall have *Philosophical heads*; or the *Philosophers* shall have *Mechanical hands*" (H 397). Sprat's explanation of the Royal Society's new model of natural philosophy insists that we view the philosopher and the "mechanic" or craftsman as engaged in related pursuits. The new philosophy, we are told, is like harvesting corn, like planing wood, like spinning thread (*H* 21, 104, 326). Philosophy, he claims, is to be understood as one of the "Arts," in the company of "*Ploughing, Gard'ning, Cookery, making Iron and Steel, Fishing, Sailing*, and many more such necessary handicrafts" (*H* 119). In the Royal Society, natural philosophy follows examples given by God himself. Turning back to the Bible, Sprat points out that the "*True God* himself" values the "*Vulgar Arts*," depicting as he does the world's "first *Monarchs* of the World, from *Adam* to *Noah*" as artisans: they teach their ancestors to "keep *Sheep*, to till the *Ground*, to plant *Vineyards*, to dwell in *Tents*, to build *Cities*, to play on the *Harp* and *Organs*, and to work in *Brass* and *Iron*" (*H*, "To the King" n.p.).

The artisan's example allowed natural philosophy to claim a particular hands-on form of experience, but the embrace of handicraft was always incomplete. As Pamela H. Smith puts it, the "new philosophers were unsettled by the involvement of the body in cognition, and they sought to control the bodily dimension of empiricism at the same time that they began to distance themselves from artisans and practitioners."[38] Bacon, even as he gives the artisan pride of place and makes scientific knowledge a "maker's knowledge," insists on the separation of artisan from philosopher, repeating Aristotle's old idea that the mechanic "does not direct his mind or stretch his hand to anything but what is useful for his task."[39] Bacon has a different idea of truth from Aristotle, but they can agree that absorption of the artisan or mechanic in the body and in the immediate task at hand prevents the broad, theoretical, or abstract thinking necessary for the philosopher. Even Sprat, whose philosophy is so close to gardening and carpentry, sticks with Bacon's point. Artisanal knowledge is productive so far as it goes, but it should not be mistaken for philosophy. In one breath Sprat highlights the "extraordinary men, though but of ordinary Trades," who are welcomed into the Royal Society; in the next, he reassures the reader that they are a tiny minority and the "farr greater Number are *Gentlemen*, free and unconfin'd" (*H* 67). Without the freedom of the gentleman (from commerce, from particular "arts" that might absorb mind and body, and from a body itself tarnished by work), philosophy cannot result, Sprat claims: "If *Mechanicks* alone were to make a Philosophy, they would bring it all into their Shops; and force it wholly to consist of Springs and Wheels, and Weights"; physicians, he claims, would have the same problem, failing to "depart farr from their Art" (*H* 66). Even in a world that privileges much about craft knowledge, we see the long-standing worry that the craftsman is too much of his craft, unable to come out of its bodily expertise for the abstract, speculative, and far-reaching knowledge that philosophy requires.

In her history of objectivity, Joanna Picciotto reads this ambivalence around the artisan as evidence of the Royal Society's attempt to bring theory and practice together in a new form of intellectual work. Scientists such as Robert Hooke, she argues, find a way to hold tight to the body's "innocence" while claiming an active, intervening labor through new instruments like the microscope. Using Adam as their "intellectual exemplar," seventeenth-century Baconians united theory with practice by drawing on Adam's innocent work, trading their own bodies for the "corporate body" his legacy offered.[40] Picciotto describes how these new practitioners of "intellectual labor" sustain this position by redefining the way knowledge works, by attempting to eliminate the distance between "knowing that" and "knowing how." Contemplation becomes "both interventionist" and "productive," offering the root of our own idea of intellectual labor.[41] But Picciotto maintains that one aspect of this older intellectual labor is superior to our practice now: Adam, that original of every man, makes intellectual labor available to all.[42]

From the outset, however, some practitioners surely would have found the Royal Society's version of this "corporate body" and inclusive intellectual labor hard to swallow. The Society's own standards, the ones both Bacon and Sprat can hardly conceal are epistemological and social, were obvious in its membership. As Michael Hunter puts it, "the statistics of elections bear out the view of the Royal Society as a high-class intellectual social club," involving many fewer merchants and tradesmen than Sprat had forecasted.[43] Some natural philosophers did cross the line and sample "lower" forms of knowledge, and some went further, putting their own hands to practical tasks. Robert Hooke, the author of *Micrographia* (1665), famously circulated through the many worlds of London, making contact with both gentleman natural philosophers and "instrument makers and skilled craftsmen."[44] Hooke made his own scientific instruments and was a surveyor and architect.[45] But he suffered for these affiliations and for his official role as the Society's Curator of Experiments, a very hands-on position. The Fellows "treated their curator as the paid employee he was,"[46] as Celina Fox puts it, and "it is doubtful whether the leisured Fellows of the Society regarded Hooke as a gentleman or even a philosopher."[47] Fox describes a related fate for Joseph Moxon, the Royal Hydrographer, who was elected to the Royal Society but who may have had votes cast against him given his earlier (and family) occupation as a printer.[48] The trouble, in other words, wasn't an individual's capacity to cross these knowledge boundaries but the social stigma of doing so.

For the gentleman Fellows of the Royal Society, the hands-on knowledge that required the removal of the body was easier on the page, perhaps, than in the laboratory. Steven Shapin describes the irony of the great chemist and experimental philosopher Robert Boyle, who distances himself from practice, even as he endorses a philosophy based on it, having assistants perform the hands-on experimentation in order to preserve his own place in "traditional gentlemanly culture." The result, as Shapin puts it, was that "Aristotle was refuted by some anonymous technicians."[49] Stephen Pumfrey finds a related split in the Enlightenment's world of public science, between the "experimental philosopher" and the professional experimenter, even though both require "head work" and "hand work." Here too, "since social status was negatively correlated with labour (specifically paid experimental labour)," there was doubt that the experimenter could also be a "philosopher."[50]

The Royal Society scientist did not have to wait until the twentieth century for the exposure of the fragility of this epistemological project. Tita Chico argues that along with Shapin and Simon Schaffer's "modest witness," literature of the seventeenth and eighteenth centuries provides "innumerable examples" of "bad scientists," dedicated to emphasizing natural philosophy's "immodesty." There are many forms of immodesty, as Chico points out, ranging from sexual to financial excess.[51] Critique took place in print and in public spectacle, in full view of a public knowledgeable about natural philosophy. Al Coppola reminds us that the character of Gimcrack in Thomas Shadwell's play *The Virtuoso* (1676) is a good index of just how public science and an awareness of scientific knowledge was at this historical moment. Gimcrack was a thinly veiled parody of Robert Hooke, and audiences (to Hooke's embarrassment) seem to have been aware of this fact. At this Restoration moment, science has gone so far into culture that a "hit play" can call on one of the "celebrated" scientists audience members knew by name.[52] The jokes Gimcrack makes also reflect an audience able to grasp the epistemological feats claimed by the new science, indeed comfortable with satire directed not just at the scientist but also at his ways of knowing.

At times, the writers in this book offer their own critiques of "modesty," but they do so on the way to presenting another way to know: through the very body that causes the disruption. Thus, while Chico and Coppola keep their focus on the natural philosopher, the context for this book requires us to turn to another aspect of Bacon's persistence in the eighteenth century in the continued visibility of the artisan and craft knowledge. Although Karl Marx identified the artisan as a figure antecedent to capitalism, the capitalist economy of eighteenth-century Britain remained structured around hand making. Richard Price explains that even in the late century, "outworking hand-labour production" was more prominent than the "machinofacture" that later capitalism calls to mind.[53] Early factories do exist during the eighteenth century in Britain, but more typical of the period are the "small-scale units of production" located in the domestic sphere and in workshops.[54] Accounts of the period have often used metaphors of "birth" to call attention to the suddenness of some forms of economic transformation (e.g., the move to paper money; the creation of the Bank of England), but economic historians now see the Industrial Revolution as a slow process and one in which "archaic and modern forms" intermingle during its gradual development.[55]

Craft was a part of this new capitalist world in another way, through the making of what Mokyr calls its "knowledge economy."[56] In this domain, the knowledge of the artisan is on greater, more public, display than ever. William Eamon has argued for the importance of "books of secrets," printed, rationalized craft practice, as an origin for experimental natural philosophy.[57] Pamela O. Long and Pamela H. Smith likewise have demonstrated that "artisanal epistemology" (Smith's term) was central to Bacon's experience-based, experimental philosophy.[58] Although these histories stop in the seventeenth century with Bacon and the natural philosophy of his followers, artisanal theorizing and rationalizing did not; in fact, it accelerated. As Celina Fox has shown, the publication of texts on "how-to" knowledge did not diminish after the Society's foundation. On the contrary, she argues, it "reached its peak in the age of Enlightenment."[59] Critically these texts and their how-to knowledge did not require the purified bodies that Picciotto and Shapin and Schaffer describe for the scientists; it was, Fox tells us, rarely even produced by those mechanic-philosopher hybrids Sprat had championed, but by practitioners of the "useful arts" outside the philosophical sphere.[60]

In this context, the exemplary late seventeenth-century text on knowledge production is not Sprat's *History* but Joseph Moxon's *Mechanick Exercises* (1677–83), a practitioner's account of the trades of the smith, the joiner, the carpenter, and the turner. Far from the natural philosopher's worries about keeping the body out of knowledge, Moxon's own concern is that the printed accounts of craft practice cannot reflect enough the body's involvement, and he describes his frustration at trying to convey that "which cannot be taught by Words."[61] When he directs the reader on individual crafts, he therefore depicts not only what happens in process but also how the craftsman positions his hands and his feet. Plates from Moxon's volume on printing show, then, not only the tools required for the trade, but how to hold them (as with the two images of hands and tools on plate 24, 1683, titled "The manner of holding the Composing Stick" and "The manner of Emptying a Stick of Letter").[62]

When the eighteenth-century "handicraft philosophers" of this book turn to questions of theory and practice, then, they do so amid a widespread knowledge of natural philosophy—including of its suspect knowledge practices—and in a world where artisanal knowledge is present and flourishing. As a result, new occasions arise for thinking—and rethinking—the relation between theory and practice. As we will see, the writers of this book do not aim to produce yet another form of "objectivity," a mode of "witnessing" or seeing that requires the removal of the body. Rather, they privilege the body's involvement in knowing and contemplate the specificity of the knowledge thus gained.

The Bodies of the Artisan

Handicraft Philosophies contributes to an important set of critical texts advancing "embodied knowledge" as a feature of the British Enlightenment, in work by Margaret Cohen, Jonathan Kramnick, and Sean Silver.[63] I too am interested in the bodily valence of knowledge in a period best known for its theory. My own book differs from these accounts, however, in tracing back embodied knowledge to its source in the culturally available discourse of natural philosophy and published how-to books on making.[64] Other accounts have grounded their philosophical histories in analogies between empiricism and later philosophy or cognitive science (or both). To take just one example, Kramnick and Silver connect early empiricism to "cognitive ecologies"—a term Edwin Hutchins locates in the 1970s—and so illuminate mind-body-world relations specified in the earlier period by way of much later ideas.[65] When Kramnick posits "ecophenomenological encounters" or Silver a "to-and-fro between models and minds," they have in mind particular models of "entanglement," formal models from later cognitive and social science.[66]

The model of the artisan is downright awkward by comparison to these seamless "to-and-fro" models that Silver and Kramnick invoke, but it also reveals some Enlightenment interests that the later models cannot.[67] First, the artisan's model is an abstract structure of relations and also a socially embedded one. In this book the artisan's embodied knowledge is not only something that happens in the workshop; its terms are published and available, ready to be used for argument. In other words, mind-body-world models are themselves being made, and the handicraft philosophers of this book are doing that making. Second, practitioner-theorists begin with a discourse in which bodies are understood as diversely "entangled" with their world from the outset. In the experiences they relate, there is no universal structure for the mind-body-world relation (as some cognitive scientific models would lead us to expect); rather, the body is already socially determined, given its materiality and perceived propensity for thought and action. After all, the mind-body-world relation is not socially neutral even in Aristotle's distribution of embodied knowledge, but rather breaks down along lines of class and gender. The Greek origin story of the craftsman in Hephaestus treads related ground, offering the craftsman's genius by way of a massive and disfigured (what we might now call disabled) body. The bodiless body of the male, gentlemanly natural philosopher only made clear to the artisan (even to the gentleman as artisan) how experience beyond this "mythical norm" might register differently.[68]

I think it's safe to say that most recent accounts of "embodied knowledge"—in any context—assume that it is a good thing. This is in large part because tactile, craft knowledge still carries the associations of its Marxist history as a model of unalienated labor.[69] Moreover, in the work of Martin Heidegger and Walter Benjamin, two of its most famous philosophical advocates, craft allows a deeper understanding of "being" and a more authentic knowledge of the world.[70] As I have already suggested, though, this moment of early capitalism leaves us much less certain of craft's redemptive qualities. At the very least, it pushes us toward Richard Sennett's conclusion that craft knowledge can have an ethical valence (as in some parts of Aristotle), but it need not do so. (Sennett's example is the atomic bomb.)[71] In the context of this book, craft knowledge can be a means to generate an ethical program, but it can also be an instrument of imperialism and power. At times, we see both valences within a single text. Joseph Banks, a visitor to Tahiti on the *Endeavour* voyage, for instance, weaponizes embodied knowledge in the service of imperial domination, even as he also uses it to consider the relation between his own epistemology and the ways of knowing he finds around him. The handicraft philosophies of this book encourage us to think about the structures of power that have determined both bodies and the knowledges they claim.[72]

The very complex formulation of a term like "cognitive ecologies," which draws on the disciplines of cognitive science, anthropology, ecological psychology (among others), also answers in advance questions that Enlightenment philosophers were keen to ask. For example, the writers in this study are quite curious about the relation between the natural world and the social one, a link newly made in other areas of philosophical inquiry in this period, most famously in the French and Scottish Enlightenments. What is the relation between the mind and society? What is the relation between the laws of nature and the laws of man? Writers from Montesquieu to Adam Ferguson to Adam Smith attempted to answer these questions for the first time. In the next section, I will explain the way that embodied knowledge, conceived through the artisan, takes its own shot at these questions, not so much in formal philosophical treatises, but in novels, in travel writing, and in autobiography.

Enlightenment authors take full advantage of the social implications of handicraft philosophies, both by using them to rejigger an account of experience that does not require a privileged or universal body, and by marshaling practical knowledge as a tool to comprehend the social world more generally. We can see this right up front in the way authors position themselves or others (including fictional others) as artisans that aren't immediately recognizable as such. Defoe writes his artisan as a thief; Mary Collier argues for her own artisan status, though her main occupation is washing clothes. This boundary-pushing use of the artisan and embodied knowledge stresses the social terms on which that knowledge rests. Defoe asks us to consider that society makes the thief's body a receptacle for the violence that society might not be willing to acknowledge as its own.[73] Collier contemplates whether her bodily exclusion from artisanal knowledge could also open up an ethical charge in its unique relation to the world.

Representing Craft

All the laborers and artisans of this project are also writers. And from the get-go natural philosophy attached its way of knowing the world to a form of representation, often referred to as the "plain stile."[74] The most famous definition of this kind of writing is Sprat's in the *History of the Royal Society*, where he describes how natural philosophical language must have a "Mathematical plainness" (recalling Aristotle's "theory") and associates such language with "the language of Artizens, Countrymen, and Merchants" (*H* 113). Blaming social schism—like that of the civil war—on bad rhetoric, Sprat calls back to a supposed earlier age "when Men deliver'd so many *Things*, almost in an equal Number of *Words*" (*H* 113). The attachment of plain writing to "purity" and certainty grounds the rhetorical efforts of many eighteenth-century genres. We find this antirhetorical rhetoric, this plainness, in the genres that have an authentic relation to what Sprat claims as the style's origin in artisans and merchants (travel writing, economic writing) but also in history writing and in the novel. Most famously in the hands of Defoe and Richardson, the plain style and the "circumstantial detail" privileged in the Royal Society allow the reader to experience fiction as certain and "true."

Yet for all the fame of this connection of words to things, in the *History* Sprat does not seem content with the reach of his claim. At times, as the *History* winds on, he is ready to leave out words entirely. He reveals his hope that actions might work in place of the words he had previously contemplated: for the Royal Society "Inventions, Motions, and Operations, will succeed in the place of Words" (*H* 327). The Society, he reiterates, is one that "prefers Works before Words" (*H* 327). From an apparent confidence that words and things could go together, that plain language could bring us to a "naked" truth, Sprat now seems to hope for something more, an artisanal making that could substitute for representation of any kind. These are his terms when he declares that the Royal Society will begin by "bringing Philosophy down again to mens sight, and practice." The new science will cast its lot not with books but with "necessary handicrafts": "the Arts of *Ploughing, Gard'ning, Cookery, making Iron and Steel, Fishing,* [and] *Sailing*" (*H* 119).

Abraham Cowley probes further the relation between language, craft, and representation in his prefatory poem to Sprat's *History*. In Cowley's "ode to the Royal Society," natural philosophy is delivered to us as a problem of representation. Cowley opens with "words" and "things" in his description of Francis Bacon:

> From Words, which are but Pictures of the Thought,
> (though We our Thoughts from them perversely drew)
> To Things, the Minds right object, he it brought,
> Like foolish birds to painted Grapes, we flew;
> He sought and gathered for our Use the tru;
> And when on Heaps the chosen Bunches lay,
> He prest them wisely the Mechanick way.
> Till all their Juyce did in one Vessel joyn,
> Ferment into a Nourishment Divine.[75]

Cowley begins by citing Bacon as the philosopher who turns us from "Words . . . / To Things, the Minds right object." Unlike Sprat, Cowley goes on to consider what kind of representation this word-thing might be, and he turns to painting, importing an elaborate simile that recalls Pliny's story of Zeuxis's painted grapes. Pliny describes a painting contest between Zeuxis and Parrhasias, in which Zeuxis's grapes appear so real that they are pecked by birds. Parrhasias, however, wins the contest, because he paints a curtain so lifelike that Zeuxis himself tries to raise it to see the painting he assumes lies beneath it. If words are like pictures in the cognitive account Cowley begins with, his simile makes "picture"—the abstraction of the first quoted line above—into a specific illusionist painting, and "we," as birds, fly to this representational triumph, certain we can peck at the grapes.

In the long life of this classical anecdote in discussions of representation, what happens next is unusual and surprising: the painterly triumph is set aside. Although Cowley recalls the painting competition, we should note that he stops it midway, giving us grapes but not the real triumph of the curtain. In Lynn Festa's reading, Cowley himself responds with his own "painted curtain of language."[76] But there is another actor in these lines, and he turns us away from language and illusionist painting entirely. While the reader is lured by the mimetic, referential grapes, Bacon, the hero of the lines, "sought and gathered" the grapes. In this turn, the reader—the "we"—is revealed not just to have been duped by a painting of grapes but by illusionist painting as the only reality we might occupy. Bacon, breaking the frame by picking the grapes, makes "use" of the illusion, choosing to press them "the Mechanick way." The *real* reality, Cowley shows us, comes from doing: the grapes are properly real only when you crush them, only when you work on them with the transformational process of a "Mechanick." Painting serves not as a guiding metaphor here but as a jumping off point for demonstrating another reality that we can grasp only when we set pictorial representation aside. Bacon's triumph recalls Jesus at the Cana wedding turning water into wine, as Cowley renders this wine "nourishment Divine." But in the context of the *History*, Jesus' miracle looks a lot like an experiment, perhaps one related to the experiment Sprat includes as an example of Royal Society work: "A Proposal for Making Wine, by Dr. Goddard." Even the move up toward divine perfection and miracle is also a movement out, toward accessible how-to knowledge and doing.

A knowledge of making places "words" in a tough spot, and Cowley and Sprat were not the only ones to notice. Any writer on craft, a knowledge grounded in the body and the hand, faces such troubles. Writing the preface to his practical manual on "handy-works," Joseph Moxon runs through the paradox for the writer on craft: rendering a knowledge of the body in words.[77] Beginning in 1677, Moxon, hydrographer royal, published *Mechanick Exercises* in monthly installments, delving into craft practice from the perspective of a practitioner. Moxon systematizes his own experience with smithing, founding, drawing, joinery, turning, engraving, printing (books and pictures), globe and map making, and making mathematical instruments. In this vast collection of practical knowledges Moxon notes right away some trouble with terminology that has affected his complete title, *Mechanick Exercises: or, The Doctrine of Handy-Works*. "Handy-works," he says, was born of rejecting the sticky term, which I have chosen for my own title, "Handy-crafts." Handicrafts, Moxon explains, "signifies Cunning, or Sleight, or Craft of the Hand," and it seems at first as though it's this relation between "craft" and "crafty"—accusations of untruth—that he wishes to avoid. But the problem with "sleight," it turns out, is that it draws attention to the "hand" you can't see in words: that which "cannot be taught by Words, but is only gained by Practice and Experience." Like any other writer of a practical handbook, Moxon will do the best he can, showing "Rules" that may be realized in experience, repetition, and use. But he insists that the text itself cannot represent the learning of the workshop, which occurs through repetition and imitation. Three hundred years later, Peter Dormer and David Pye begin their seminal accounts of craft by making the same claim: craft knowledge is "not easily described by language" (Dormer); "workmanship" cannot "be conveyed in words and by drawing" (Pye).[78] All three of these writers, over centuries, begin books on craft with something like, "You know what you're going to read here? That's not really *it*." In all cases, they assert bodily action and knowledge over writing, acknowledging that the conventions of writing—and perhaps words themselves—do not allow the craftsman to specify the expertise and knowledge in his body; the know-how of the hand may never truly be constituted as knowledge available only to the head.

The body's knowledge has not been thought to have much importance for the way we read texts of the Enlightenment. Take *realism*, the umbrella term under which the literary use of "plain style" usually falls. Discussions of eighteenth-century realism usually stop at Cowley's account of Zeuxis's grapes without acknowledging his punchline: the transformative "art" of the georgic Bacon. Svetlana Alpers's account of the relation of empiricist philosophy to Dutch still-life painting sometimes serves as analog for descriptive writing; this is how Shapin and Schaffer adopt it in their account of the "circumstantial detail" involved in scientific reporting.[79] Perhaps, critics have hypothesized, the detail-obsessed prose writing of the period might work the same way. Maximillian Novak compares Defoe's realism to painting, and Mary Baine Campbell connects *Pamela*'s realism to the detail of Hooke's engravings in *Micrographia*.[80] Such accounts seem to bring literary and scientific texts together in the realm of "virtual witnessing," as Shapin and Schaffer put it, with plain language creating a nearly seamless passage from one visual experience of an event to the reader's imagined experience of the same.[81]

In a way that's rarely acknowledged, however, the "plain style" can be strangely capacious.[82] To grasp this we need to set aside the assumption that novelistic realism is based on a visual model. Let me offer a brief example from *Robinson Crusoe*, the go-to text for discussions of the period's realistic prose. On the island Crusoe is not shy about his own role as a maker, performing a series of what he calls "experiments": making pots, weaving baskets, and planting corn. He offers a map of the weather that is quite close to the weather chart of Hooke's that Sprat includes as an example of Royal Society reporting in his *History*. To be sure, as Ilse Vickers has argued, Defoe is an entrenched proponent of Royal Society natural philosophy.[83] We can see the influence of this philosophy in the experience-based reporting Crusoe gives us in his description of how he makes a wooden mortar and pestle:

> so after a great deal of Time lost in searching for a Stone, I gave it over, and resolv'd to look out for a great Block of hard Wood, which I found indeed much easier, and getting one as big as I had Strength to stir, I rounded it, and form'd it in the Out-side with my Axe and Hatchet, and then with the Help of Fire, and infinite Labour, made a hollow Place in it, as the *Indians* in *Brasil* make their canoes.[84]

If we cast ourselves as the "modest witness" of Crusoe's experiment, we find him searching for particular materials (hard wood, as opposed to soft), specifying his tools, and giving us a process that we might be able to replicate: "forming" the mortar on the outside with an "Axe and Hatchet" and making a "hollow Place in it" with fire. It's not a long recipe, but it is one.

This is not quite all the passage gives us, though, and Peter Dear's account—a more elaborate account of Royal Society form than Shapin and Schaffer's—allows us to see more about the forms of natural philosophical experience contained here. The most obvious sort of experience pertains to Crusoe's record of an "experience," as Dear puts it, a contrived and singular "event" registered at one point in time, by a particular person, in the plain language for which Defoe is well known.[85] This is the hallmark of the new philosophy, its attention not just to what had always been but to what was observable in this single case. *Robinson Crusoe* can't get enough of day-to-day reporting, presenting dated journal entries, even if the weather or Crusoe's actions repeat over time. Thus, we're fairly sure that what Crusoe relates to us in the recipe for the mortar and pestle is what happened *this* time. Yet there are other elements that don't seem to cohere easily with this description. He "rounded" and he "form'd"—here, we are within a single event—but then he slips in that he has worked with "infinite Labour."[86] After this passage Crusoe lists the steps of his experiments to bake bread after grinding the grain with his new tools, ending with "several Cakes of the Rice, and Puddings."[87] Here again, we think we know what's going on with the event of making. But Crusoe follows up that description with the declaration that "all these Things took me up most Part of the third Year of my Abode here."[88] In both cases, Defoe forces the reader to do a double-take at the end of the event, when we realize that there was much more experience—repetition, shared labor, trial and error, "infinite" work—contained in the account than its language had at first let on.

Crusoe's account of experiment contains two quite different forms of experience: that of the singular event and its materials; and that of working and reworking in practice for the real thing—the "practice" and "experience" of Moxon's description. This latter kind of experience is the sort that Defoe and his readers would have associated with craft knowledge: it is embodied, it is repetitious, it is learning by doing. We find, too, in the passage on the mortar and pestle, that Crusoe, sometimes conceived as an isolated Royal Society scientist, has gathered knowledge from a number of sources, as when he remarks that he uses as his model how "the *Indians* in *Brasil* make their canoes." Lying behind the European imperialist's use of nature to create civilization is indigenous knowledge that Crusoe has absorbed and that he now imitates. We see something similar—the breadth of knowledge and its establishment in community—when, in making his baskets, Crusoe calls back to the basketweavers he observed as a child, who let him "lend[] a hand" to the work (78). Crusoe's Royal Society experiments, it turns out, contain layers of practical knowledge from a variety of sources, some of them well beyond the conscious individual experience that we might assume drives his narrative.

I touch down on this single experiment in Robinson Crusoe to show how realism can quickly exceed the objects and events that are usually ascribed to it. Defoe's accounting for experience pushes a kind of extra experience—repetition, practice—back into the single story visible in the event of experiment. In so doing, he gives the reader something much like what Cowley does in bringing us to a particular plane of representation, only to disrupt it and to show us other varieties of experience, those bound to the body and to community. As Sprat suspects, plain language can't contain experimental processual knowledge very well. Cowley and Defoe dramatize this by focusing on a referential way of knowing, convincing us that this is all we see, and then showing us more experience, different in kind. For the writers of realism who privilege the processual qualities of craft knowledge, trompe l'oeil realism is merely a starting point.

Knowing Society

The British Enlightenment's contribution to our later "social science" is generally thought to come through its adoption of scientific thought for the study of humankind. Indeed, if the seventeenth century is the source for the "scientific revolution," the eighteenth century is credited with taking this revolution to the human world. The later period is full of philosophers attempting to test the possibility that human laws might be as discoverable as natural laws. Montesquieu begins with this hope as fact: "Laws, in their most general signification, are the necessary relations arising from the nature of things. In this sense all beings have their laws: the Deity has his laws, the material world its laws, the intelligences superior to man have their laws, the beasts their laws, man his laws."[89] Hume, also writing midcentury, makes clear that the same impulse drives his own very different philosophical project: the "science of man."[90] This "science of man" is generally thought to have its endpoint in the disciplines of sociology and anthropology, formed—and professionalized—in the nineteenth century.[91]

In *Handicraft Philosophies*, I describe moments when practitioners make use of natural philosophy to describe the human world around them—and find it wanting. To be sure, there are plenty of instances where things seem to go right for the natural philosopher on the ground. We might look to travel writing in this regard, where William Dampier's best-selling accounts of his travels attend to natural specimens right alongside the arrowheads or practices of the "Moskito Indians" he encounters.[92] Dampier's work would be picked up by the Royal Society and his example would be part and parcel of the recommendations that the Society would give to future travelers[93] Even by the 1770s, when Cook and Banks receive directions from the Royal Society for the *Endeavour* voyage, the letter from the Society's president encourages them to treat "governments," systems of communication, and natural land formations as the same kind of object.[94] In retrospect, it is easy to see Banks and other writers stuck in the world David Carrithers describes: "the key Enlightenment fallacy," he claims, "was the conviction that observing individuals in society required an approach no different from what Newton had employed in observing nature."[95]

On the voyage, however, Banks's ways of knowing become increasingly complex, especially after the ship arrives in Tahiti. We witness this in

Banks's journal, kept separate from his explicitly scientific writings, where he chronicles scientific discoveries alongside the perceptual apparatuses they require. And when he turns to the objects of Tahitian culture—cloth, tattoo patterns, and so on—he explicitly works to present these objects as distinct from the objects of the natural world, using craft knowledge to do so. Margaret T. Hodgen has written that in the seventeenth century travelers and natural philosophers generally had no use for "Manners and Customs"—those practices that would become central for later anthropology and sociology.[96] But, a century later, this is not the case for Banks, Royal Society Fellow and gentleman, both.

Handicraft Philosophies is not a book on travel writing, though I do feature two famous eighteenth-century travels (Cook and Banks's voyage to Oceania and Johnson and Boswell's trip to Scotland) at its center. In so doing, I mean to show that travel writing that has long been considered "philosophical" (Johnson's) is not so far in its concerns from that considered "scientific" (Banks's). The other chapters in the book contextualize these encounters with other societies, however, showing how texts in diverse genres—poetry, the novel, autobiography—also endorse embodied knowledge as a way to know something about "customs and manners." As the book unfolds, I show how arguments with different starting points (an account of a child thief, a felt exclusion from a georgic poem on work, the definition of beauty) end by using the body to grasp "social knowledge," as I call this distinctive epistemological endpoint in the book's title.

As we can see from Mr. Spectator's example, society can be an especially hard object to know with standard natural philosophical tools. Indeed Addison's claim isn't necessarily that society creates such spectator trouble, which already exists for the spectators of the Royal Society, but that society as an object lays bare these very strange conditions for knowing. In the realm of bodily knowledge, in the practices of the artisan figure, there is no sticking with modesty, of course, since the goal in such knowledge making is to be near, to practice, to imitate, to gain another body's knowledge. One reason that scientific remove might not seem easily applicable to society is established by Mr. Spectator: society invites you in, does not give you an easy place to be a "looker-on." Addison shows us that this can be understood as a relation between theory and practice, and craft knowledge—in bringing together the two—thus seems a better fit for the complex object of study that society is.

The legacy of "handicraft philosophies," then, is not in the nineteenth century's embrace of an objective study of cultures but in what comes after it, in structuralist and poststructural approaches to anthropology. We can see twentieth-century anthropological theory continually rework its object of study in a way that recalls these eighteenth-century beginnings. Gregory Bateson compares culture to other "objects" studied by science: "be it an animal, a plant, or a community."[97] But this view does not last. Clifford Geertz points, instead, to culture as a text, and to the process of studying it as a "hermeneutics."[98] And Michael Jackson insists on an "anthropology of the body," pointed against "the semiotic model of culture" and toward (again, though differently) "the world of materiality and biology."[99] The history of anthropological theory, then, is very much one of grappling with culture in terms of ontology: What kind of thing is this? And how must it be understood?

Some of the most trenchant critiques of anthropology as a natural science have drawn on theory and practice to make their claims. In the context of handicraft philosophy, it is hard to ignore Pierre Bourdieu's title *Outline of a Theory of Practice*, or even his earlier *Craft of Sociology*, both of which loudly declare Bourdieu's intervention as a rigorous reevaluation of the relation between, as he puts it "theoretical knowledge" and "practical knowledge." Indeed, when he turns toward his "theory of practice," he writes over and against structuralism by moving toward a relationship between individual practice, habitus, and structure.[100] We hear an echo of handicraft philosophy too when at the beginning of Michel de Certeau's *The Practice of Everyday Life* the philosopher delves into the "logic of . . . practices" by way of asking, "What is an art or 'way of making'"?[101] These much later answers to the question of how one is to know society counter the persistence of scientific objectivism (and other sorts of distance) by making good on Enlightenment writers' intuition that "practical knowledge" has something special to say about "practices" or "techniques," as well as about the structures or patterns through which they may be known.

Handicraft Philosophies argues that students of society have long turned to theory and practice as a way to grasp their object of knowledge, and that, for at least as long, they have positioned their new ways of knowing over and against the one proposed by the natural scientist. The following pages describe some of these early epistemological experiments.

Chapter Descriptions

I frame each chapter with a keyword that illuminates craft's cultural relevance for eighteenth-century Britain: education, labor, art, science, use, and the fetish. All of these terms track the way handicraft pulls together social and epistemological questions.

My chapters and texts reach across social positions of authorship and across genres, in chapters that treat novels, poetry, philosophy, travel narrative, and autobiography. In considering issues of representation, I use this generic variety to emphasize both the many textual locations for "handicraft philosophy" (outside formal treatises) and the way that individual sets of generic rules provide the grounds for different experiments with craft's relation to experience. Beyond genre, craft knowledge necessarily poses questions about the adequacy of writing as representation, something I track in Banks's and Hogarth's turn to a combination of text and visual image.

Chapter 1, "Education: Crime, Punishment, and Realism in Defoe," examines the relevance of craft knowledge for Defoe's account of education in his novel *Colonel Jack*. Here, Defoe, following Baconian educationalists, uses craft knowledge to account for experience beyond the reasonable, conscious self. The artisan's method of education, of course, already does this: it is based on imitation in the workshop, which operates through the hand, or the movement of the body, rather than through the book or mind. Defoe comprehends this bodily realm of experience as important not just for prescribing studies but for understanding the society of which the educated person is a part. I begin with Defoe's *The Complete English Tradesman* and *The Complete English Gentleman*, both of which offer models for education as a way to comprehend the new economic world driven by trade. Turning away from abstract models of system like mercantilism's circulation, Defoe uses craft to comprehend how the tradesman is situated in this rapidly changing society.

In his 1722 novel, *Colonel Jack*, Defoe brings craft to the social events of crime and punishment in Jack's London world, where he examines the figure of the artisan as a child thief. In his fiction Defoe turns loose the model of education on the world in general, for Jack has no tutor for his experience-based learning. Over the course of the novel, Defoe offers the reader various experiments with society's relation to the individual, forcing us to discriminate between natural and social knowledge and to account for the ways in which the world works on persons, both through and beyond Locke's understanding of the individual person.

Defoe uses craft knowledge to understand society in part by expanding who the artisan might be: not just a carpenter and a natural philosopher, but also a thief. In Chapter 2, "Labor: Craft Poetry in Duck and Collier," the poets Stephen Duck and Mary Collier put additional pressure on the definition of the artisan, interrogating the definition of her or his labor. Writing in the georgic mode, Duck and Collier claim the form of artisanal knowledge in Virgil's original poem: husbandry. But both are eager to redefine the terms for the Virgilian poet, whose labor has an uncertain relation to that of the husbandman. Writing back to the georgic intellectual laborer of the Royal Society, Duck and Collier claim for poetry a connection to embodied work, even as they also use craft labor to push against the new economic pressures on the laborer.

In the second part of the chapter, I turn to Mary Collier's response to Duck. Collier embraces Duck's account of embodied knowledge but pushes it further. For one thing, Collier, a washerwoman, quarrels with the category of artisanal knowledge that could include husbandry but not her own work. What, she forces us to ask, is the definition of "handicraft" that could underwrite intellectual labor? And might not poetry itself partake of some of the most unseemly, embodied aspects of craft, even those that do not seem especially productive? Through craft, through knowledge that is embodied, Collier urges us to see that there is more than one form of embodiment, since the body that does the knowing is itself the product of social norms and expectations. Collier probes both the injustice and the potential ethics of the "care" that defines feminine labor.

Chapter 3, "Art: Handicraft Lines in Hogarth" pulls questions surrounding the practical side of "art" from the work of georgic poets into formal aesthetics. Hogarth is the most straightforward "handicraft philosopher" of this book, engaged in writing a theory of beauty, *The Analysis of Beauty*, from the standpoint of a practitioner. In fact, he allows us to see his work as a kind of "History of Trades" project in the Royal Society tradition. In this chapter Hogarth defends an older, capacious "art" against the academies' "fine art," and he does so by insisting on a form of experience available only to the artisan. Pulling back technique and materiality into accounts of shape and line,

Hogarth's artisan folds more than one reality into a form. Across the line of beauty, Hogarth's account of utility, and, finally, his reflections on action, we find a group of experiments that paradoxically use a universal line to spell out the way form can contain the reality of society as the reality of its own making.

The chapters of the book thus far have examined the way a popular understanding of experimental philosophy intersects with craft knowledge in writings by figures associated with literature and fine art. In Chapter 4, "Science: Practicing Science in Banks's Tahiti," I turn directly to a scientist aboard Captain Cook's *Endeavour* voyage, the botanist, Joseph Banks. On the face of it, Banks would seem an unlikely candidate for "handicraft philosophies." Not only is he a gentleman and a Fellow of the Royal Society; he is also a devotee of Linnaeus, the famous taxonomist, whose method categorizes plants and animals through a process of abstraction. I begin with Banks's examination of the science practiced onboard the *Endeavour*, most centrally Cook's cartography and Banks's own Linnaean botany. In his journal Banks is preoccupied with the representational methods attached to each: the creation of forms on paper or canvas to represent islands on a grid or specimens in a system. Banks's writing takes each method and examines its material, craft roots: in the imagine on the page and canvas, in the bodies that write and experiment. In a voyage famous for its scientific images, Banks pulls representation toward the body and environment, away from abstraction and easy replication. He does so, however, not as Defoe did, by opposing natural philosophy or science to craft, but by opposing certain forms of scientific representation to the craft knowledge of experiment.

In the final section of the chapter, I turn to Banks's small sketches from the journal and their relation to the plain style of Royal Society reporting, arguing that, like Hogarth, Banks uses craft knowledge to specify the particularity of social knowledge, even as he also attempts to formulate ties to indigenous Tahitian society based on "practice."

Chapter 5, "Use: Useless Bodies in Johnson and Boswell," examines how travel writing pushes ideas of making into "use." Although Johnson is usually considered a "philosophical" traveler and Banks a "scientific" one, we find the great literary man contemplating a related set of questions about the ability of experience-based knowledge—here, specifically antiquarianism—to register his experience of Scotland. Craft appears in this chapter in a new guise, in terms of "use" and the "useful" in Johnson's consistent, colonialist accusation that the Scots are bad makers. Even as he tries to one-up the Scots with his own knowledge of making, craft turns back on the substance of Johnson's journal, producing an account of the bodily basis for such things as measurement and darkness.

In the second half of the chapter, I turn to Boswell's journal from the same expedition, to show how he takes up craft knowledge in his own terms. Here, I argue, Boswell turns back Johnson's account of embodied knowledge on the body of the great man himself. In figuring Johnson's own body as useless, Boswell performs a kind of anti-imperial gesture, even as he paves the way for his more neutral, gentlemanly body to serve as the standard for participatory observation.

In Chapter 6, "Fetish: Equiano's Antislavery Craft," Olaudah Equiano takes the African fetish as the grounds for redefining craft making as this book has considered it. The African fetish is a European concept, meant to show the "otherness" of the African by asserting as a logical mistake the making of an object that then serves as a deity. Equiano marshals this supposedly mistaken account of making to a different end, showing the way it makes analogous African and European ways of thinking around the mysterious kinds of power that are located in made objects and the surprising relations of those objects to the body. In his hands, then, the problem of the fetish becomes a way of thinking about the origins of all kinds of social practices, belief among them.

The Interesting Narrative details Equiano's embodiment as a craft practitioner, especially in his identity as a navigator. Equiano battles out the problems that craft knowledge, embodied knowledge, can pose for the enslaved person. In this book's earlier examples, bodily knowledge is assumed to be something that belongs to you; indeed, you can have this knowledge even if you are a poor boy in London or a woman who washes clothes—even, that is, if you are kept from formal education and property ownership. But the enslaved person is different: he does not own his own body. For this very reason, antislavery works often focus on the slave's mind or soul; if the body is not his own, personhood need not be located there. Equiano, however, takes a different tack, embracing craft but with a knowledge of the problem Defoe shows us in *Colonel Jack:* in the economy of slavery, craft knowledge is the superior knowledge of the master visited on the body of the enslaved. In response, Equiano turns away from the "mastery" we usually associate with the artisan, only to then then draw on the artisanal workshop and its collectivity to show us a form of subjectivity that is present across bodies. Equiano's advocates not a universal subjectivity knowable as an abstraction but a shared, embodied subjectivity knowable across differences.

We know from its use in imperialism that craft is not simply emancipatory. Equiano shows us a contemporary consideration of this epistemological plight. Craft, he argues, must call our attention to the body that does the work, to the social construction of that body, and to the forms of knowledge permitted to it. Equiano's use of craft is hopeful, precisely because it brings to the fore how conceptions of freedom and agency rely on our sense of how we make the world—and how we are made by it.

ONE

Education

CRIME, PUNISHMENT, AND REALISM IN DEFOE

DANIEL DEFOE'S FICTIONS DISPLAY HIS strategic use of natural philosophy's antirhetorical rhetoric, the plain language that Thomas Sprat hoped might draw words toward things. *Robinson Crusoe*'s lists show the "concrete particularity" that results,[1] as when Crusoe describes one load of things from the shipwreck: "Bread, Rice, three Dutch Cheeses, five Pieces of dry'd Goat's Flesh . . . and a little Remainder of European Corn . . . as for Liquors, I found several Cases of Bottles belonging to our Skipper, in which were some Cordial Waters, and in all about five or six Gallons of Rack."[2] The world of *Robinson Crusoe* is full of things: a Wall "thickned with Pieces of Timber, old Cables," and containing "seven little Holes"; a "delicious Vale" full of "Abundance of Cocoa Tees, Orange, and Lemon, and Citron Trees" with some "green limes" used to make a drink; weapons described as "two Fowling-Pieces" with "Swan-Shot, as big as small Pistol Bullets" and "four Muskets" loaded with "two Slugs, and five small Bullets each."[3]

This chapter examines how Defoe refines an object-oriented realism in the interest of a new epistemological project: a knowledge of society.[4] In the Introduction I discussed how focusing on objects, even in *Robinson Crusoe*, can lead us to miss Defoe's experiments with the way that "practical" or embodied knowledge is related to the world he describes. I continue the analysis here, taking as my primary text a slightly later, and certainly lesser-known, novel, *Colonel Jack* (1722). Part of a group of what are called Defoe's "criminal novels," *Colonel Jack* does not fit easily into the "formal realism" account of prose, though its writing remains detailed.[5] Looking at how Defoe works through and around a plain language model in this different text allows us to recognize his efforts to formulate a knowledge of society through examining how it intersects with, and ultimately bears on, the individual's body.

In *Colonel Jack* the handicraft philosopher is Jack himself, a young boy living on the streets of London, who turns to pickpocketing and larger crimes. I track two different ways in which natural philosophy is brought to bear on Defoe's social experiment: through his ideas of Baconian educationalist theory and through his experimental conceptions of crime and punishment. In both cases, Defoe shows us how a knowledge of society depends on using the artisan to reconfigure natural philosophy's model of knowledge: its ideas of observation and the observer, and the concept of experience that grounds it.

The Artisan-Tradesman

Defoe was in a good position to formulate the tradesman as a relative of Sprat's craftsman-philosopher. Defoe, from a Presbyterian family, was educated at the Dissenting Academy at Newington Green. The academy was led by Charles Morton, who was "intimately acquainted with the research of experimental scientists of his day" and was the author of the *Compendium Physicae* (written c. 1680) which he completed during Defoe's years at the academy (and a copy of which Defoe likely retained long thereafter).[6] The *Compendium*, which included citations and discussions of Galileo, Kepler, Newton, Harvey, and Boyle, among others, would become the standard natural sciences text at Harvard.[7] Morton's classrooms contained air pumps and scientific instruments; Defoe's education, in addition to experimental philosophy, included modern languages, geography, and astronomy. Defoe would have known Royal Society science and may even have performed experiments in one of Morton's labs. It is important for what follows that he came to experimental science as the son of a well-respected tallow chandler (candle maker), a craftsman and member of one of the powerful London guilds.[8] For this reason, Defoe may have been especially attentive to the strain of craft knowledge that moves through natural philosophy's account of knowledge.

We see Morton's style of education on display in Defoe's *Complete English Tradesman*.[9] The first volume of Defoe's *Tradesman* is addressed on its title page to "young beginners" and is sometimes called a "conduct book" for the tradesman.[10] In it, Defoe champions the experience-based style of education he experienced at Newington Green. Defoe's career—and his own self-conception—was marked by insults from gentlemen of the literary world, and he clearly has Jonathan Swift and others in mind when he defends his education and claims the tradesman's knowledge to be superior for the modern world. He puts it this way in the *Review*, writing of the "Merchant":

> [H]is Learning Excels the meer Scholar in Greek and Latin, as much as that does the Illiterate Person, that cannot Write or Read: He Understands Languages without Books; Geography without Maps; his Journals and Trading-Voyages delineate the World; his Foreign Exchanges, Protests and Procurations, speak all Tongues; he sits in his Counting-House and Converses with all Nations, and keeps up the most exquisite and extensive part of human Society in a Universal Correspondence.[11]

In the *Tradesman*, Defoe lays out the methods behind this superior learning, advising the tradesman to bypass books and go straight to the world for education.

By the time Morton taught Defoe at Newington Green, educational theory influenced by Bacon was on the rise. By the middle of the eighteenth century, it was "a fully codifiable educational program."[12] The title of *The Complete English Tradesman* proclaims its status as a practical handbook of education in the tradition of Henry Peacham's *The Compleat Gentleman* (1622).[13] In both the *Tradesman* and Defoe's later *The Complete English Gentleman* (1726), Defoe takes aim at Peacham's restatement of the general assumption that gentlemen are born with virtues that require a particular educational approach; Defoe's substitution of "tradesman" for "gentleman" in his first title is itself an affront to such aristocratic ideology. Arguing for a kind of education available to any mind, Defoe follows the Baconian educationalists, most famous among them John Locke in his *Some Thoughts Concerning Education* (1693).[14] Locke's advice is directed to the gentleman on the education of his son, but Defoe sees here—as have other readers—the potential for a program of education that could extend much deeper into society's ranks. Indeed, Defoe has experienced the reach of such a program himself.

Locke takes up the Baconian program by emphasizing education that occurs through immediate sensory experience. Locke's child does not begin with books. As the philosopher observes in the *Essay Concerning Human Understanding*, the child's mind is particular: it is "pretty late, before most Children get *Ideas* of the Operations of their own Minds," and for years they operate without reason and through "outward Sensations" only.[15] When Locke turns to education, then, he begins with these sensations, instructing parents not to dress children too warmly, to fit them with leaky shoes, and to regularly bathe their feet in cold water. In such passages Locke uses his medical training to have us think about education in terms of the body. When scholars have acknowledged this, they have usually understood Locke to be taking aim at the body in order to prepare the mind.[16] But Locke's claim is stronger than that, and even later education for the boy or man leads to bodily engagement. Throughout the treatise, Locke is attentive to how the body creates knowledge: whether the "manly Thoughts" that come from dancing, or the perfect characters that come from a well-held pen.[17] Even when gentlemanly status has been achieved, Locke says, that accomplished gentleman should not relinquish practical, bodily knowledge. Rather, "I would have him *learn a Trade, a Manual Trade*," Locke declares; the man might be a "*Carpenter, Joyner,* or *Turner*" (*E* 257). Locke sees such manual activity—such handiwork—as promoting a mastery of the self, a constant exercise in "*submit[ting] his Appetite to Reason*" (*E* 255). Locke says "reason" here, but of course this is not the way we customarily think of the intellectual faculty of reason, for this craft work, this handiwork, occurs through the body's practice, as well as through the mind.

We have already seen Defoe's endorsement of experience-based learning for the artisan-tradesman of his first educational treatise. We can see there too how Defoe tends to take an idea of Locke's and push it further, toward its limit. Locke endorses book learning that begins with more immediate experience. For instance, geography should start with the globe, which could be comprehended as "an exercise of the Eyes and Memory" (*E* 235). From there, the student might tackle "*Arithmetick*," "the first sort of abstract Reasoning," until he is "Master of the Art of Numbers" (*E* 235). The student could take those numbers back to geography, to longitude and latitude, and the comprehension of maps, the figure and position of the constellations. Finally, the tutor could present "a Draught of the *Copernican* System, and therein explain to him the Situation of the Planets, their respective Distances from the Sun, the Center of their Revolutions" (*E* 236). Look at the way Defoe intervenes just here, early in the *Tradesman*, when he tells us of a particular philosopher "excellently skill'd in the noble science or study of astronomy" (*CET* 62). This philosopher, also a tutor, runs into trouble in his lesson when he finds that he cannot convey to his students the sun's movement on its axis. He searches for a "simily" (*CET* 62) but finds something better:

> [B]y accident he saw his maid *Betty* trundling her mop: surpris'd with the exactness of the motion to describe the thing he wanted, he goes into his study, calls his pupils about him, and tells them that *Betty*, who herself knew nothing of the matter, could shew them the sun revolving about itself in a more lively manner than ever he could. Accordingly *Betty* was call'd, and bad bring out her mop, when placing his scholars in a due position, opposite not to the face of the maid, but to her left side, so that they could see the end of the mop, when it whirl'd round upon her arm; they took it immediately; there was the broad headed nail in the center, which was as the body of the sun, and the thrums whisking round, flinging the water about every way by innumerable little streams, describing exactly the rays of the sun darting light from the center to the whole system. (*CET* 63)

Like Morton, and as Locke advises, Defoe's philosopher teaches modern natural philosophy, something in which the universities were notoriously deficient. Here, he highlights Galileo's discovery of the sun's rotation on its axis and the assumptions of the Copernican system with the sun's rays "darting light from the center." But intervening in Locke's program, Defoe suggests that at the culmination of Locke's very gradual introduction of abstract knowledge, we must resort, again, to the material: to the mop's physical motions. Defoe also accentuates just how bodily the learning process of this materialized abstraction must be: the students must stand to Betty's "left side," achieving what he calls "due position" for the experience-based learning to take.

The *Tradesman*, however, does not merely follow Bacon and Locke on education. In addition, Defoe uses the treatise to make a strong claim for the kind of knowledge that the tradesman assumes. This goes beyond his claims for the tradesman as worthwhile, as the social equal of the gentleman. We recognize his terms when he opposes the tradesman to the scholars he calls out as "mechanicks"—they are Bacon's and Sprat's terms for practical-theoretical knowledge. Indeed, Defoe sets up the tradesman not just as a practical learner but as an ideal combination of practice and theory, placing him in league with—if not above—the natural philosopher. Sprat has told us that the Royal Society is a "*union* of *eyes*, and *hands*," and Defoe follows with the characterization of the tradesman "application both of his hands and head to his business" (*CET* 98).[18]

Defoe begins the *Tradesman* by defining the tradesman over and against the artisan. After settling on the definition of "tradesman" in "*England*, and especially in *London*, and the South part of *Britain*," he observes:

> all sorts of warehousekeepers, shopkeepers, whether wholesale dealers, or retailers of goods, are called *tradesmen*; or to explain it by another word *trading men*: such are, whether wholesale or retale, our *grocers, mercers, linen* and *woollen drapers*, Blackwell-hall *factors, tobacconists, haberdashers*, whether of hats or small wares, *glovers, hosiers, milleners, booksellers, stationers*, and all other shopkeepers, who do not actually work upon, make, or manufacture the goods they sell. (*CET* 35)

The tradesman is defined against those who make "*handicrafts*, such as *smiths, shoemakers, founders, joiners, carpenters, carvers, turners*, and the like" (*CET* 35). Like the philosopher, the tradesman has a larger purview, for he is not concerned with his trade only, but with "all the inland trade of *England*" (*CET* 37). He must not know only a single trade but comprehend many.

Such comprehension sounds a great deal like Sprat's philosopher's turn toward, then away from, the artisan. The natural philosopher must visit the artisan to know how things are made and thus to gain a knowledge of the natural world. Defoe's tradesman needs this knowledge too, and in a manner that Bacon would have endorsed, for he must know the way things are made in order to gauge their value in the marketplace:

> If you go to *Warwickshire* to buy cheese, you demand the cheese of the first make, because that is the best. If you go to *Suffolk* to buy butter, you refuse the butter of the first make, because that is not the best, but you bargain for the right rowing butter, which is the butter that is made when the cows are turn'd into the grounds where the grass has been mow'd, and the hay carried off, and grown again, and in so many other cases. (*CET* 57)

Defoe describes this under the heading of the "languages of trade," but rather than on terminology the focus here is on the technical knowledge signaled by such "language" (*CET* 57). The butter example, as Defoe includes it here, emphasizes that we must know not just the right name of the butter to ask for but also the state of the land and the position of the cow that could make such milk. Defoe calls this knowledge "these lower things"; such "lower things" tell us how the product is made, when and how the work is done (*CET* 57).

As with Sprat's natural philosopher, Defoe's theory-savvy tradesman remains close—if not identical to—the craftsman. Defoe's most consistent analog for the tradesman is the navigator, that notorious synthesizer of theory and practice of ancient times, a practitioner of what we would now call applied science.[19] When Defoe lays out rules for successful trade, he draws further on ancient examples that highlight craft. For example, at the beginning of Letter IV, he begins by disagreeing with Horace's statement in his second epistle "that no man ought to go from one business to another . . . *Tractent fabrilia, fabri*, Every man to his own business" (*CET* 58).[20] We recall that this ability to grasp multiple businesses is basic to Defoe's definition of the tradesman that he presents at the outset. But what Defoe offers us here—"Every man to his own business"—is a pretty loose gloss of Horace's words, which in the Loeb translation are rendered, "carpenters handle carpenters' tools."[21] As much as Defoe wishes to separate the tradesman from the carpenter's "handicraft," then, his disagreement with Horace involves conflating the two, rendering carpentry as "business." Even when Defoe gives us examples of the multifaceted tradesman, his examples track with tradesman-makers: a drysalter and a scarlet-dyer. Both tradesmen are scientist craftsmen, involved in chemistry and manufacture. (An eighteenth-century dry-salter dealt in chemicals such as varnish and dye, in addition to salt for preserving foods.) In keeping craft close, Defoe stresses that the tradesman's knowledge should not be too abstract; he should be able to "turn his hand to this or that trade" (*CET* 58); acquiring the kind of embodied learning that, to take Horace's examples, a navigator, physician, or carpenter, would need.

As in Sprat's treatise, then, a broader sort of philosophical knowledge comes back to making and the maker. The tradesman is positioned against the artisan, but he must have a knowledge that is like the craft knowledge of the workshop or the factory. Following Sprat, Defoe claims, too, that the tradesman's "stile" must live up to his actions, advocating for what he calls the "trading stile."[22] In his explanation Defoe seems sometimes to be making a simple point about writing directly and clearly, so that transactions do not "end[] in loss both of money and credit" (*CET* 52). The stakes for clarity are high in the world of trade. But in the term "trading stile" Defoe shows us that he understands Sprat and others have claimed the tradesman as an origin for the new natural philosophy. And his account of this antirhetorical rhetoric is in keeping with Sprat's: the plain or trading stile is "a free unconstrain'd way of speaking" (*CET* 47), over and against "*Tropes and Figures,*" "amplifications, digressions and swellings of style" (Sprat);[23] and the "flourish[]" of the "poet" (Defoe; *CET* 47).

The tradesman's style of learning and representing falls in step with the natural philosopher's, and thus shows how the man of commerce possesses the same form of knowledge as the philosopher. In this sense, Defoe both extends Sprat and corrects the earlier writer's hesitancy about extending philosophy beyond the gentleman. The tradesman learns as the artisan does, through practice and repetition. In the shop or in the warehouse, Defoe says, "he sees how other men go on, and there he learns how to go on himself" (*CET* 61). "Going on" is a rather vague account of this activity, but Defoe gets more specific—and more material—as he continues. Take conversation: "as writing teaches to write, *scribendo discis scribere,* so conversing among tradesmen will make him a tradesman" (*CET* 60). Conversation is learned by doing, and it is hard to imagine it any other way. But what is startling about Defoe's example here is that the account of conversation takes writing as the vehicle of explanation. In pushing verbal language back into the written word, Defoe cannily expands writing itself to doing—to "going on." Defoe had already made this point in accounting for the "Trading Stile" as "useful." But here he thickens our sense of what this use might mean: not just able to function in the world but also attached to the body, the product of *hand*writing. Defoe continues to move written words toward manual practice when he argues in favor of knowing the "terms of art" for each trade (*CET* 57). He explains, "to be able to speak or write to any particular handicraft or manufacturer in his own dialect . . . is as necessary as it is for a seaman to understand the names of all the several things belonging to a ship" (*CET* 54). The tradesman's multiple languages are like names for tools: not just plain but tied to function and use.[24] If Sprat worried about the mechanic's philosophy of springs and wheels, Defoe seems unconcerned about the metaphor that makes the metalanguage of the tradesman just a set of languages, tools he has acquired through knowledge of various trades. Hands and head, body and mind, the tradesman makes good on practice-as-theory, leaning toward the body in a way Sprat's gentleman could not do. The tradesman, as Defoe puts it, must bring his "full attention of the mind, and full attendance of the person" to the task (*CET* 67).

But the tradesman's focus exceeds the natural world. He may know how cheese and bricks are made, possessing the knowledge that Bacon wanted reflected in his histories of trades, but his domain is also the social world. His experience is not just of natural forces but of social ones, of conversation. This is commonplace in a text like Locke's of course, which shows us how the body itself is molded by (carefully chosen) social forces; this is how good manners are inculcated. Defoe goes one step further, however, wrapping that social knowledge—the knowledge not just of how materials behave but how people do—into the larger philosophical claims about theory and practice.

In one of his most surprising turns, Defoe brings the tradesman's practical knowledge directly into contact with representation, in ways that exceed what we are led to expect from the "plain" or "trading stile." The *Tradesman*, if discussed in terms of Defoe's economic theories, generally shows him championing the tools of the new economic world, tracking "circulation" and advocating for credit as essential to an expanding economy. In all of this, Defoe promotes a how-to account of keeping the books, which he says is essential for the tradesman. In his long account of double-entry bookkeeping Defoe calls up the form that stands at the center of Mary Poovey's history of the "modern fact," a history that, like Defoe's *Tradesman*, attempts to correct the source of the "fact" in science and place it back in trade.[25] Indeed, whereas other historians have taken philosophers' own accounts of the "deracinated particular" as foundational for the fact, Poovey instead sees a philosopher like Bacon as the inheritor of a complex formal account of facticity developed in trade, whose "epistemological issues carry over into the sciences."[26] In Poovey's analysis, the number-based fact comes into focus, not as a theory-free particular, but as one utterly dependent on a structure for its meaning: the symmetry and order of the double-entry system creates the fact by instantiating the formal means by which that fact can be verified and the tradesman's reputation preserved. This theory-bound fact is present even in Bacon's philosophy, where, Poovey argues, particulars operate with an "aura of theory."[27]

In the *Supplement to the Complete English Tradesman* (1727), Defoe offers "Direction" to the tradesman on keeping books.[28] He also offers sample pages of accounting books from the double-entry system (cash book, petty cash book, day book or journal, and ledger), following the tradition that Poovey charts from the fifteenth century onward. In each case, Defoe introduces the circumstances behind the new sort of book, then follows his sample pages with explanation and analysis, breaking down particular entries by stressing aspects of formatting and interweaving real-world stories about account books, as when he narrates how a dead tradesman's book proves he did not die in debt as his family expected (*CET* 256). Throughout, Defoe stresses the "care" and "pains" attached to this accounting method, and he stresses its importance for proving the tradesman's practice in a court of law, as well as allowing him to keep track of the work of his subordinates (*CET* 30, 262). Such authority and transparency are the hallmarks of the double-entry method.

Defoe's double-entry method, however, is different in one critical respect from the classic method outlined by Poovey. As Sandra Sherman has written, Defoe eliminates the formal balance that is the prime feature of the standard method. Rather than having the accountant enter the "double" of "double entry," the same sum listed as debit and credit, Defoe sends the accountant to the actual money on hand. As Sherman puts it, "In classical theory, balance does not refer to what is *actually* in the cash chest, determined by physical counting, but to an agreement among discursive categories from which one can derive the state of one's cash." Defoe's accounting, by contrast, is "transgressive," in that it "mov[es] the ground of affirmation from Text to World."[29] Defoe notes the way that "the bag and the book . . . [are] brought together": "by casting and re-casting up, telling, and telling over and over again the money" (*CET* 215). For Sherman, this signals a destabilization of accounting, signals the "fragility of the text," and moves accounting toward credit.[30]

With Poovey's account of facticity in mind, however, we can see another end to Defoe's depiction of counting. Defoe has us witness the physical casting and telling—"over and over again"—stressing the repetition that sutures text to world, book to bag. Defoe gives us this relation of body to system as an answer to Poovey's theory-based facticity, the way in which truth is rendered visible through formal, textual parallelism. The "world" undoes truth (as Sherman worries) only if we are stuck in a world comprised only of texts. Defoe assures us that we are not. Rather, he shows us how to move from one domain to another, injecting sensory experience—motions of the hand—where it does not belong, collapsing the visual form of the double entry.

Defoe binds form to experience earlier, too, when he shows the way accounts could be done entirely through manual, tactile work. Earlier in the text, Defoe tells the story of a man who could not read and who "knew nothing of figures" and yet excelled in trade (*CET* 209). To keep accounts, "He made notches upon sticks for all the middling sums, and scor'd with chalk for lesser things; he had drawers for every particular customer's name, which his memory supplied, for he knew every particular drawer, tho' he had a great many" (*CET* 208). Defoe compares the man's organizing marks to "*Egyptian* Hieroglyphicks," stressing the marks as both symbols and language (*CET* 208). Organization here is also the work of material objects: in a drawer "was nothing but little pieces of split sticks, like laths, with chalk-marks on them . . . every stick had notches on one side for single pounds, on the other side for tens of pounds, and so higher" (*CET* 208–9). When the man has to calculate accounts, he uses a method involving six spoons: "By this he told up to six; if he had any occasion to tell any farther, he began again, as we do after the number ten in our ordinary numeration" (*CET* 209). With these physical counters, the man is able to calculate any sum: "by the strength of his head he cou'd number as many more as he pleased, multiplying them always by six's, but never higher" (*CET* 209). Most important for Defoe's purpose, the process of keeping accounts is here translated into concrete objects and materials, made tactile and described as made, crafted objects. In comparing the sticks to "laths" (thin pieces of wood that could underlie tile work), Defoe puts us in mind of the making associated with his own work as a tile manufacturer. He also asks us to think about the technical terms of a given trade as tools.

What at first seems a whimsical example of an unusual practice can be understood, in light of Poovey's analysis, as another push on the very structure that produces the "deracinated particular" in the first place. In this description, Defoe materializes system: again, we are reminded of Sprat's nightmare of gears and wheels as philosophy. More than that, these tools are themselves handmade. Economic form should illuminate the modern world of which it is a part, showing us its truth. We might even see economics as a system that operates beyond the individual, as in Adam Smith's "impartial spectator." Defoe, by contrast, keeps reintegrating theory into practice, making it hard to sort out abstraction from craft. The tradesman's knowledge, he suggests here, is of the social world, of the world that his knowledge is meant to represent.

The Artisan-Tradesman-Thief

The *Tradesman*'s theory of trade-as-craft is embodied by the young thief Colonel Jack, from Defoe's 1722 novel by that name. *Colonel Jack* is a novel of education and as such has a home with the forms of educational theory Defoe develops in the *Tradesman* and in his more overt educational treatise, *The Complete English Gentleman*.[31] Defoe's setup for *Colonel Jack* indicates that he wishes us to pay attention to the way children develop in the world: he places at its early center two different boys named Jack, raised and nursed by the same woman, one of whom has a lower-class origin (the Captain) and one of whom is the illegitimate son of a gentleman (the Colonel).[32] The Colonel does fare much better in life, but Defoe leads us to question whether or not blood alone can be held responsible, because he gifts his hero with something else: the education of Locke and Morton.

Colonel Jack is in some ways the perfect Lockean student. For a poor boy on the London streets, there is no need to manufacture the conditions for physical resiliency that Locke imagines for the gentleman's son: shoes probably leak, if there are shoes at all. David Blewett assumes that Jack's lack of education entails a "dismal fate," but Locke's and Defoe's accounts of what education looks like can lead us to conclude otherwise.[33] Reaching for Locke's most challenging examples for the tutor, Defoe turns to history early in *Colonel Jack*. History is tough for Locke's program, because it can't easily be learned outside of books; thus, Locke suggests that the student first master "Geography" and "*Chronology*," beginning with a physical globe, before turning to and trying to "retain[]" history (*E* 237). Locke's student takes a long road: From chronology he turns to classical history; then to Tully, Puffendorf, and Grotius on the founding of society; and, finally, to "the *Law* of his Country" found in the constitution and government, and the "ancient Books of the *Common Law*" (*E* 239, 240).

Colonel Jack shortcuts this laborious list and arrives at contemporary history through chats with "Seamen and Soldiers," with whom he talks "about the great Sea-Fights, or Battles on Shore, that any of them had been in."[34] Jack reassures the reader that he has been just fine without chronology: "I could soon, that is to say, in a few Years, give almost as good an Account of the *Dutch* War, and of the Fights at Sea, the Battles in *Flanders*, the taking of *Maestricht*, and the like, as any of those that had been there," as well as "the

Wars in *Oliver's* time, the Death of King *Charles* the first, and the like" (*CJ* 68). Defoe does not leave us in any doubt about the kind of knowledge this is. "I was," the older Jack relates, "a kind of an Historian, and tho' I had read no Books, and never had any Books to read, yet I cou'd give a tollerable Account of what had been done, and of what was then a doing in the World, especially in those things that our own People were concern'd in" (*CJ* 68). Here Defoe turns Locke's own advice back on the philosopher's account of history. Just as Locke recommended that Latin be learned by mothers reading the Bible to their children, so Defoe shows us a similar oral process for learning history, one that additionally masters knowledge by hearing from the historical actors themselves. Like Locke's ideal student, Jack does not privilege books, even in the subject most obviously confined to book-learning—after all, he can't afford books and, at this point in the novel, he doesn't know how to read. Like Defoe himself, Jack is "no scholar" but is educated for the world in which he will live, which will include military battles of the likes of what he records here.

If Jack's self-education brings in the world as tutor, he moves even further into Defoe's world of experience-based learning in his first professional role as a pickpocket. For here, beginning his young life of crime, Jack assumes the artisan-tradesman identity that Defoe would specify in the *Tradesman*. Defoe describes Jack's pickpocketing as a "Trade" and "business" (*CJ* 70), but he is especially plain that theft is the "Art of Thieving," its actions "Feats of Art," the work of an "Artist" (*CJ* 99, 143, 107). We recall Defoe's command that the tradesman bring "hand and head," a combination that is required too for picking pockets. Pickpocketing demands physical action and thus must be learned by watching and doing, just as one might learn in a workshop: Jack, Defoe tells us "go[es] Apprentice" to his trade, is "bred up" to it (*CJ* 76). The trade involves "Dexterity" and "Skill," the very terms in which we would expect craft work to be evaluated (indeed, Locke evaluates the gentleman's handicrafts using just these words) (*CJ* 98, 99). In this period, pickpocketing was understood both as a beginner's crime, a low rung on the ladder of criminal activity, and one that could demonstrate dazzling skill and expertise.[35]

In the account of Jack's first solo crime, Defoe works out further his terms for the pickpocket's craft. The criminals proceed to the Royal Exchange, the center of London commerce since the sixteenth century. The boys "hanker'd about in *Castle-Alley*, and in *Swithins-alley*, and at the Coffee-house-doors"

(*CJ* 96). Defoe leads us through the streets surrounding the Exchange, and, as with many of the London scenes in the novel, we can plot their movements on a map. The boys are not terribly successful the first time around, but they soon return for a profitable venture:

> Here we began to act separately, and I undertook to Walk by my self, and the first thing I did accurately, was a trick I play'd, that argued some Skill, for a new Beginner, for I had never seen any Business of that Kind done before: I saw two Gentlemen mighty Eager in Talk, and one pull'd out a Pocket-book two or three times, and then slipt it into his Coat-pocket again, and then out it came again, and Papers were taken out, and others put in; and then in it went again, and so several times, the Man being still warmly Engag'd with another Man, and two or three others standing hard by them; the last time he put his Pocket-book into his Pocket, he might have been said, to thro' it in, rather than put it in with his Hand, and the Book lay End way, resting upon some other Book, or something else in his Pocket; so that it did not go quite down, but one Corner of it was seen above his Pocket. (*CJ* 99)

Although we are not yet to the crime, we see Jack's boast about his own "Skill," and in what follows, we are allowed to follow his thoughts and movements.

> When seeing the Book pass, and repass, into the Pocket, and out of the Pocket, as above, it came immediately into my Head, certainly, I might get that Pocket-book out, if I were Nimble, and I warrant *Will* would have it, if he saw it go and come, to and again, as I did: But when I saw it Hang by the way, as I have said; Now, 'tis mine said I, to my self, and crossing the Alley, I brush'd smoothly but closely by the Man, with my Hand down flat to my own Side, and taking hold of it by the Corner that appear'd; the Book came so light into my Hand, it was impossible the Gentleman should feel the least motion, or any body else see me take it away. (*CJ* 99–100)

In an important way, Defoe is not describing anything all that exceptional here. A man leaves his wallet hanging out of his pocket, and a thief, at the ready, snatches it easily. But we are drawn to see both the wily intelligence and keen observation of the thief (head) as well as the lightning-fast, trained movements of his hand: "it was impossible the Gentleman should feel the least motion," Jack tells us. The pocketbook "came so light into [Jack's] Hand."

Defoe gives us a practical-theoretical actor and a mode of representation to match, though the latter can be hard to see based on the extant definitions of Defoe's "realism." On the one hand, street names in London feel like a familiar kind of referential detail, the kind that makes up Ian Watt's formal realism or constitute Roland Barthes's "reality effect."[36] This is the detail of reference and of excess, the kind of thing that has Crusoe listing how many cheeses he brings from the ship, with whole paragraphs devoted to such items. Critics interested in Defoe's "realism" have tended to focus on these details.[37] Sometimes they have portrayed Defoe as a realist painter, alongside Zeuxis and Parrhasias, as when Maximillian E. Novak writes that "Defoe often tried to achieve the vividness of the painter."[38] Such claims about realism have returned in more recent times, with writers on the literature and history of science relating Hooke's microscopy—particularly, its detailed illustrations, its "minute particulars"—to the details of *Pamela* or *Tristram Shandy*.[39]

To be sure, not every writer on eighteenth-century realism has rested here. Dorothy Van Ghent, for her part, finds that Defoe's passages would offer poor directions for a painter. Van Ghent argues that it is things, not description, that Defoe is after: "In saying that the world of *Moll Flanders* is made up to a large extent of *things*, we do not mean that it is world rich in physical, sensuous textures—in images for the eye or for the tactile sense or for the tongue or the ear or for the sense of temperature or the sense of pressure. It is extraordinarily barren of such images."[40] Van Ghent's point of comparison is the nineteenth-century novel, which is rich in just this experiential detail. Cynthia Sundberg Wall, also grappling with how there can be so many things and so little "richness," argues that Defoe's passages are "unvisualized" but rich, suggesting that the kind of circumstantial detail we don't see would have been present for their eighteenth-century readers, would could read life's details into the immediacy of "description."[41]

In all of these (even opposing) understandings of Defoe's realism, we find little that can account for the pocketbook passage's sort of specificity. After all, there is no painterly detail here. What do the men at the exchange look like? What does the pocketbook look like? Defoe does not seem concerned with such questions. But Van Ghent and Wall don't seem to meet the mark either, as they mainly show us what Defoe leaves out. We might look again at the long part of the first passage:

> [O]ne pull'd out a Pocket-book two or three times, and then slipt it into his Coat-pocket again, and then out it came again, and Papers were taken out, and others put in; and then in it went again, and so several times, the Man being still warmly Engag'd with another Man, and two or three others standing hard by them; the last time he put his Pocket-book into his Pocket, he might have been said, to thro' it in, rather than put it in with his Hand, and the Book lay End way, resting upon some other Book, or something else in his Pocket; so that it did not go quite down, but one Corner of it was seen above his Pocket. (*CJ* 99)

Van Ghent and Wall, while rightly pointing out that Defoe is not the committed realist that could bend to an analogy with Dutch painting, seem committed to sensory detail as the marker of realist success. Defoe, for his part, seems interested in a feature of objects that does not get much play in talk of representing the real: movement. Defoe's account draws our attention to the back-and-forth of the objects' movements, and, indeed, less to the object itself than to how it lies. We could turn back to Cowley's lines, when he writes in the prefatory poem to Sprat's *History*, "The real Object must command / Each Judgment of his Eye, and Motion of his Hand."[42] Svetlana Alpers (whose book forms the basis of many a realism assessment in this period) takes from these lines a restatement of Hooke's connection between sight and the artist's hand, but Cowley's later lines on Bacon the mechanic crushing grapes suggest where Sprat and the Royal Society are headed, and where Defoe goes too.[43] Defoe draws out reality as movement in the passage on the pocketbook, our eyes able to track the arc of the pocketbook and its relation to the man's hand and to his pocket. He has us do this multiple times, repeating the back-and-forth words in the passage, so that the experience is conveyed in these repetitions.

Accounts of literary realism have not had much care for process and action, but Defoe's models were hardly limited to classical rhetoric.[44] The passage from *Colonel Jack* resonates more fully with Peter Dear's account of literary form within early Royal Society reports of experimental events.[45] Dear describes how the "actual, discrete event" of the experiment is represented with two kinds of detail: the "circumstantial detail" that sets up and frames the experiment and the details of the event or action of the experiment itself.[46] The "circumstantial detail" of this account is most like the "concrete particularity" often associated with the novel: it contains particularity, the use of referential language, and a focus on objects. In the history of science, Robert Boyle is famous for this brand of detail, often running, as Dear puts it, "to excruciating lengths." Such details could appear in the dating of experiments, in its preparations, and they signal the "good faith employed by the virtuoso."[47]

As opposed to the models for detail often employed by literary critics, Dear's model of Royal Society reporting includes two different kinds of detail that do not work against each other, but instead support the same experimental description. Let me pause to consider Dear's example from Boyle before turning back to Defoe's prose. Dear takes Boyle's Torricellian experiment and Boyle's "circumstantial" details of the setup: "We took a slender and very curiously blown cylinder of glass, of near three foot in length, and whose bore had in diameter a quarter of an inch, wanting a hair's breadth," Boyle begins, before proceeding to describe the pipe: "hermetically sealed at one end [it] was, at the other, filled with quicksilver, care being taken in the filling, that as few bubbles as was possible should be left in the mercury."[48] But Dear separates this kind of particularity from what he finds in the "'experience' itself," from the "report of what happened."[49] Setup procedures could be repeated but the experimental event with the air pump happens this way, on this occasion:

> But when after this, the feathers being placed as before, we repeated the experiment by carefully pumping out the air, neither I nor any of the bystanders could perceive anything of turning in the descent of the feathers; and yet for further security we let them fall twice more in the unexhausted receiver, and found them to turn in falling as before; whereas when we did a third time let them fall in the well exhausted receiver, they fell after the same manner as they had done formerly.[50]

The force of Dear's distinction is that, although there may be particulars in this description, they are not the focus of the passage. Boyle's event is concerned with actions and movements. The details specify the experimenter's own movements and the movements of the feathers in the "unexhausted receiver." Boyle reports as an active participant in the scene and leads the reader through the steps, not rushing to conclude, so that she may "virtually witness" for herself what has happened.[51]

If we think of Defoe's scene as a recorded experiment, Royal Society style, we see that the street names possess one kind of detail and the motions taken by the pocketbook another. Like Boyle watching the feathers, we watch the movements of the pocketbook, even if we are not sure what they add up to. And Boyle highlights too the way that Defoe's pocketbook works like a "body" in a physics experiment, its movements the focus, even over and above the man's actions. (The feathers are not interesting in themselves in that experiment, though their motions may show us something about the air's resistance.) We begin with a person in charge of a deliberate action, "one pull'd out a Pocket-book two or three times, and then slipt it into his Coat-pocket again." But as the action is repeated, the pocketbook takes on a life of its own, leaving aside the mind of the man: "then out it came again, and Papers were taken out, and others put in, and then in it went again, and so several times." Finally, "the Book lay End way . . . so that it did not go quite down." Defoe prepares us for the crime by describing its object in terms of bodies and forces.

I will return to the pocketbook, but let me treat, briefly, another scene in which Defoe has the same kind of focus on action.[52] In this later scene we are led to a very particular location for Will and Jack's crime: the market in West-Smithfield, on a Friday. An "antient Country Gentleman" has sold some steers and receives the money:

> having some of it in a Bag, and the Bag in his Hand, he was taken with a sudden fit of Coughing, and stands to Cough, resting his Hand with the Bag of Money in it, upon a Bulk-head of a Shop, just by the *Cloyster-Gate* in *Smithfield, that is to say,* within three or four Doors of it; we were both just behind him, says *Will* to me, stand ready; upon this, he makes an artificial stumble, and falls with his Head just against the old Gentleman in the very Moment, when he was Coughing ready to be strangl'd, and quite Spent for want of Breath.
>
> The violence of the blow beat the old Gentleman quite down, the Bag of Money did not immediately fly out of his Hand, but I run to get hold of it, and gave it a quick snatch, pulled it clean away, and run like the Wind . . . (*CJ* 109)

Again, Defoe's detail seems excessive and again it moves us in a way that a realism of visual detail would not lead us to expect. In this passage, one thief throws himself against an old man, and the other thief grabs the bag of money from the man's hand. But this is not how Defoe has us conceive of the action. He asks us, rather, to trace the relations between bodies and forces, so much so that this tracing dominates our reading of the passage. The body of the gentleman convulses with coughing and "in this very moment," Will's own body intervenes. Here, the passage builds on the earlier one, showing us not just motion but relative forces. We are shown the position of Will's head "just against the old Gentleman" and the force of the movement, the "blow" to the other body. Defoe has us gauge too what this force is enough to accomplish—it "beat the old Gentleman quite down"—and yet it is insufficient: "the Bag of Money did not immediately fly out of his Hand." Crime is reduced to bodies at work on each other.

Defoe's passages are engrossing, and it's easy to forget what exactly we're examining. After all, Defoe has used a form from natural philosophical reporting to capture acts that are both economic and social. In both cases, thieves of lesser means take money or valuable objects from gentlemen. The pocketbook theft gives Will and Jack the most explicit claim to being fellow tradesmen, because the boys use their knowledge of the different kinds of bills and receipts found there to leverage their value. Ultimately, Jack negotiates with the gentlemen victims of the crime, pretending to be an outsider who can sell the item back to them. These men, it turns out, are perfectly willing to negotiate in this manner. They are also appreciative of something else: the pickpocket's skill, referring to the thief as an "artist" and making clear that they would like an account of how the theft was done. In this, Defoe wraps the crime into just another commercial negotiation, even as he depicts the pickpocket as having artisan-like craft secrets that those in a higher social position are keen to know.

The gentlemen, it turns out, are perceptive readers of the natural-scientific model we have been given, for they see the thief as a kind of social artisan. In giving us crimes as experiments, Defoe seems to play out—as is his wont, to an extravagant degree—the basic notion current in this period that social laws might be visible through the same processes as natural laws. Montesquieu banks on this in *The Spirit of Laws*, when he attempts to figure out such laws for human societies, but the connection need not be made only in the realm of the formal philosophical treatise; an author trained in natural philosophy with a keen interest in the social world might come to the kind of inventive small-scale application that Defoe does. If we read the experiments this way, we see Defoe engaging the larger question of what kind of thing society is (another natural world? something different?). And we see him aware that criminal activity might allow us a window into that world. We can see such an interest in the mere writing of the "criminal" fictions, of course, but here Defoe puts the point with stunning specificity and on a microlevel.

We might be tempted to say, though, that this doesn't get Defoe terribly far. After all, doesn't his examination of society just reduce it to nature, to the push and pull of bodies? Defoe wonders something like this—aloud—in the earlier example of the pocketbook theft. In that earlier passage Defoe separates out the initial set of observations (where the book passes and repasses, finally ending at an angle in the pocket) from the crime, the snatching of the book, that comes next. In between, Jack breaks to offer this analysis:

> This Careless way of Men putting their Pocket-books into a Coat-pocket, which is so easily Div'd into, by the least Boy that has been us'd to the Trade, can never be too much blam'd; the Gentlemen are in great Hurries, their Heads and Thoughts entirely taken up, and it is impossible they should be Guarded enough against such little Hawks Eyed Creatures, as we were; and therefore, they ought either never to put their Pocket-books up at all, or to put them up more secure, or to put nothing of Value into them. (*CJ* 99)

We stop here for what seems like a moral of the story. Perhaps we readers might profit from the boys' crime, Defoe suggests, avoiding such things in our own lives. This repeats a familiar move from Defoe's criminal novels: an assurance to the reader that we are being instructed, not merely delighted, by criminal behavior.

Even as we absorb this moral, however, we might feel a small shock from an explanation that is so out of keeping with the representational frame Defoe has offered us as readers. Defoe trains our attention on the pocketbook's movements, but it turns out that this experiment-level reporting has not told us what we needed to know about the object of our perception. Experiment decrees that this is a one-time-only viewing, reflected in the immediacy of the prose description. We find out now, however, that these movements are not mere forces in the natural world; they are undergirded by habit or custom, which determines a "Careless way of Men" that can be identified and attached to a class of persons, gentlemen. After the moralizing interjection, we realize that we are looking at a different kind of movement than the experimental form allowed us to see.

Jack's analysis, then, pushes us to read beyond the "surface" that is often set up for realism and to grasp the various kinds of experience that reside in and beyond form. In her account of the eighteenth-century novel, Helen Thompson pulls apart "'realist' reference" from "mimetic imitation," showing how "eighteenth-century novels make explicit the *production* of empirical reality as the reader's encounter with forms and powers that enable sensational knowledge."[53] Thompson's unit is the corpuscle; she is interested in Defoe's attempt to represent that which cannot be seen. We see a related interrogation of vision here, though pushed into the realm of actions that are plainly available to sight.

Defoe's introduction of this habitual, social knowledge is paired with a turn toward irony, then an expansion of our ways of considering observation. For although this account of the pocketbook has just been presented to us as "the report of what happened" (in Dear's terms), Defoe draws our attention to the perceiver of the action—and how that matters for what we have seen. Jack calls himself one of the "little Hawks Eyed Creatures," which might suggest that this is a natural way of seeing. But he is quick to indicate that the "Hawks" are those "us'd to the Trade." He sees well, because he is practiced and taught; he has, as we can say with Defoe's earlier metaphors in mind, acquired this specialized sight as craft. Specialized observers, of course, are well in keeping with experiment; Jack's record of the movements of the object are so precise, because, like Sprat's philosophers, he is a "plain, diligent, and laborious observer[]" (*H* 72).

At this point, we might be tempted to rest with "irony" as an explanation of Defoe's substitution of the young thief for the trained Royal Society observer.[54] But "irony" does not tell us much, especially when we continue to hover over a topic about which we know Defoe was quite serious: practical knowledge. Defoe certainly seems to poke fun at the convention of the gentleman observer here (much as he pokes fun at the gentleman figure in the novel and in *The Complete English Gentleman*). But this has important implications for perception. Tita Chico and Al Coppola both describe impassioned observers in satires of the period, "immodest" observers who take down the modest scientist—his modesty, after all, just a matter of gender and class privilege.[55] Defoe's child thief certainly can be classified as an "immodest" observer. Indeed Defoe teasingly recalls Royal Society method in his thieves' casual conversation. Jack says of the Captain that "he scarse ever Pitch'd upon any thing in his Eye, but he carried it off with his Hands" (*CJ* 98–99). If Sprat draws our attention to the natural philosophical "*union* of eyes, and *hands*" (*H* 85; and Cowley: "Each Judgment of his Eye, and Motion of his Hand"), Defoe has those hands feel and acquire, both, shifting disinterest to the most interested move of all: theft.

Defoe, however, does not allow us to stop there, because he has drawn for us a very particular idea of what constitutes theft. The thief's role in the taking of the pocketbook and the bag of money configures his own actions carefully and depicts for us a kind of making. If we move back to Bacon and think of the craftsman as desirable because he routinely experiments with natural forces, getting to know them by interacting with them, bringing them under his control, we see Jack perform a related work. He describes the act of theft to us as entering a web of forces: "it came immediately into my Head, certainly, I might get that Pocket-book out, if I were Nimble, and I warrant *Will* would have it, if he saw it go and come, to and again, as I did." The thief might have intervened just in the back-and-forth of the book, Jack tells us, but he doesn't need to try for quite such a spectacular feat: "But when I saw it Hang by the way, as I have said; Now, 'tis mine said I, to my self, and crossing the Alley, I brush'd smoothly but closely by the Man, with my Hand down flat to my own Side, and taking hold of it by the Corner that appear'd" (*CJ* 100). Like the artisan, the thief observes, knows the forces in play, and then intervenes, making good on craft knowledge as "productive" knowledge.

Jack observes, then, but he also intervenes. In modest witnessing, the witness hardly pauses to note where his hand was, or how his body interacts with the object. This embodied knowing, this form of participation, of interacting with the object, is the domain of the artisan that, in the main, gets excluded from reports of experimentation. We can see, then, that Defoe leads us through one kind of perceptual experience, experiment, only to turn it twice: toward the social custom of the gentleman and toward the embodied, participatory knowledge of the thief. We should read this as part of Defoe's career-long concern with education, with how we know what we know. We have seen his straightforward accounts of experience-based learning and how he brings those into the novel through Jack's early experiences. Here, he expands our view of what this might entail, because the thief offers a particular sort of perception—and action—that seems uniquely fitted to the social world. The thief sees something about the gentlemen that they cannot see: their manners, their patterned social actions. What natural philosophy can't do lets us see what Jack can and how his participatory, body-based knowledge might not just be a witty answer to Locke or Sprat but a way to access knowledge about the social world that molds those around him.

When Defoe has us trade the modest gentleman observer for the boy-thief observer in this passage, he encourages us to ask what about the criminal might make him a different kind of ideal observer. Critics have long realized that criminals in Defoe's fictions work out questions of "social organization" or point to "societal character."[56] This may be because of the kind of social relations that criminals embody. On the one hand, Defoe makes a big deal of Jack as a product of "necessity," as formed by the society—the poverty, the sailors, the criminal friends—around him. Education makes the man. On the other hand, though, the criminal does not have the kind of "embedded" identity that could be forecasted for one of Locke's gentlemen students, for example, who are made by the habits acquired or instilled in them, because the criminal is at odds with the society that makes him. Indeed, as Lincoln Faller points out, eighteenth-century criminals are viewed as "defiant disturbers of the social order," even actors who lie outside the foundational social contract on which that political order depends.[57]

Defoe is notoriously kind to and humanizing of his criminal characters. But, as Faller says, those characters (Jack, Moll, Roxana, Singleton) "never quite get absorbed into the social; but then neither do they ever seem so incoherent as to stand insolubly beyond it."[58] This in-between constitutes a different kind of privileged position for observation than the one the modest witness sets up. The pickpocket observes patterns of behavior; he must constitute a social world of manners in order to intervene in the right place at the right time. He must know how men work here or there: what drives them but also what their unconscious, absent-minded actions look like. He can do this as a member of society who is also not exactly of it, a gentleman-in-training who imitates but also takes stock of the ways in which imitation works—how it catches, how it comes to define a set of people. This is ordinary behavior for Defoe's criminals, but it is also, as he draws out here, an extraordinary contribution to ideas about how knowing the social world might be different from knowing the natural world. You might need to be inside and outside, an observer and a participant both, in order to get to the crux of what's really happening.

The Body of the Artisan-Tradesman-Criminal

Through Jack, Defoe encourages us to see that the criminal possesses a kind of self-consciousness or awareness about his own relation to society and about the way in which the pieces of society that he inhabits—from his odd, liminal position—fit together. At the same time, Jack is formed by society quite directly, much more so than the student Locke imagines in *Education*. Having no books and needing no books, having and needing no tutor, Jack goes straight to the seamen and sailors.

This seems like a blessing, an intense closing in on the source of knowledge. But it's worth reflecting that such a thing would not be possible in Locke's world, and not just because he isn't imaginative enough to grab the possibility. For Locke children's minds come with extraordinary limitations: most important, children do not yet possess the faculty of reason. If thrown into the world, they would be unable to discriminate between its influences. The fact that Jack talks to sailors and comes out of it with "history"—rather than debauchery—seems, on such an account, like pure luck. When you are taught, experience-based education is guided strictly by the person in charge; if that person is a thief, then you'll be trained in thievery too.

If one reads the *Essay Concerning Human Understanding*, then *Education*, Locke, we can see, comes to the matter of an education by habit and custom only because he must. In the *Essay*, habit-driven knowledge is bad; we recall Locke's discussion of associationism and its perversions of knowledge. Experience, especially in the form of habit or custom, works around the conscious mind; that is both its power and its danger. When Locke turns to education, he embraces a feature of experience that the *Essay* wanted to reject, but he accepts it for practical reasons. After all, if children cannot reason, what other way would there be to access their minds? A tutor's tight control over this powerful form of experience called habit could be a powerful tool in setting a child on the right path.

Locke's most famous metaphor for consciousness is the blank slate on which experience might be written. In *Education*, however, he turns away from this writing metaphor and toward material substances that can be "worked on." Craft making is common in educational treatises before Locke, too, even in Peacham. Although Locke rejects Peacham's premise that there is something inherently different and superior about gentlemen's minds, he follows the earlier writer in considering the mind something that the educator may "form" (*E* 80). We see this in his account of the child's mind as "Wax, to be moulded and fashioned as one pleases" (*E* 265). This tactile appraisal continues through the work; Locke writes early that the child's will is the site of "Suppleness" and later that minds are "tender" but may "harden" (*E* 111, 103, 180).

If Defoe has us see Jack as an artisan, Locke reminds us that education based on experience is only possible in the first place because the mind is "tender" or "pliant" (103), able to be worked on. That is, he reminds us that the child is more crafted object than craftsman. The child's mind and body begin with classical accounts of wax, but in *Education* Locke follows Bacon's move to the craftsman. No longer is wax simply touched and shaped by fire; in these more elaborate metaphors, Locke has us think in terms of making: habits are "woven into the very Principles of his Nature," he writes, and, just a few paragraphs later, he describes habits as "wrought into the Mind" (*E* 110, 111). Wrought here means "worked into," and its eighteenth-century meanings extend through multiple kinds of artisanal work: spinning wool, welding metal.[59] Habit works as a form of skilled making. This emphasis is present too in what is sometimes seen as a separate metaphor in the treatise: agriculture. Husbandry and cultivation put us in mind of the georgic's skilled, productive labor, whether shaping metal or the earth.

Throughout the novel, Jack is formed by circumstances or "necessity." His poverty makes him a thief, and he perpetually falls into conditions that range from slavery to war. But the language of craft—of habit "wrought into" the mind—allows us to see how Defoe works with other forms of experience that do their making beyond the overt plot or situations of the novel. Let's turn to an example of realism that seems to mimic the plain style of reporting that we saw in the thefts of pocketbook and moneybag. By this point in the novel the two Jacks, Colonel and Captain, have fled England and have thieved their way to Scotland. Nearly as soon as they arrive in Edinburgh, the two aim to "see the Town" and find themselves in Mercart-Cross, "where we saw a great Parade or kind of Meeting." Captain Jack feels excited at the potential for criminal acts, but the boys are "surpriz'd with a Sight":

> [W]e observ'd the People running on a sudden, as to see some strange Thing just coming along, and strange it was indeed; for we see two Men naked from the Wast upwards, run by us as swift as the Wind, and we imagin'd nothing, but that it was two Men running a Race for some mighty Wager; on a sudden, we found two long small Ropes or Lines, which hung down at first pull'd straight, and the two Racers stopp'd, and stood still, one close by the other; we could not imagine what this meant, but the Reader may judge at our surprize when we found a Man follow after, who had the ends of both those Lines in his Hands, and who, when he came up to them, gave each of them two frightful Lashes with a Wire-whip, or Lash, which he held in the other Hand; and then the two poor naked Wretches run on again to the length of their Line or Tether, where they waited for the like Salutation; and in this manner they Danc'd the length of the whole Street, which is about half a Mile. (*CJ* 149)

Much as in the earlier passages that mimic natural philosophical experiment, this one uses the "plain" style ironically—not to tell us simply and transparently what is in front of us but to force us to question what we are looking at. Defoe does this most obviously with the speed at which the passage proceeds. Although it is hard to imagine more than a few seconds—or even a second in total—between these stages of mentally processing the scene, Defoe writes a passage that takes several times that length to read, with Jack leading us through the stages of perception and preliminary conclusions, so that we experience the "surprise" he does in the appearance of the man with the whip.

In this slow relation, the object we are looking at is transformed before us. Along with Jack we see first the crowd only and "some strange Thing." Following this, "two Men naked from the Wast upwards" are made visible and termed "Racers." Finally, the "two long Ropes or lines" that hold them are indicated, along with the man who follows them and "gave each of them a Lash." The passage presents the reader with set of images, leaving us to add it all up as we process and conceptualize along with Jack. We are faced with a kind of puzzle, a "strange Thing" that we resolve, piece by piece, into a scene of punishment.

The scene, however, stubbornly refuses to add up, getting stuck in the middle, at the point when the boys see "Ropes" but are unable to reconceptualize the action and still refer to the men as "Racers." Finally, when we are presented with the man with the whip, we expect this categorization to change, for Jack to sum up (as I just did) that what we are looking at is punishment, the men criminals.[60] Instead, Jack offers this: "and then the two poor naked Wretches run on again to the length of their Line or Tether, where they waited for the like Salutation; and in this manner they Danc'd the length of the whole Street, which is about half a Mile." Notice that the earlier (incorrect) conceptualization of the racer is not replaced here. Instead, confusingly, Jack opts for metaphors, describing the whipping as "Salutation" and the movements of the men as dancing. That is, in the moment of potential conceptual resolution for the reader—*this* is what we're looking at!—Jack confusingly gives us more actions that are similar in kind to the physical movements he sees.

What are we looking at? Or, to use the language of the boys, what does it *mean*? ("We could not imagine what that meant," they say at the appearance of the ropes.) Defoe seems to offer two answers to this, one consonant with what we have seen so far and the other new. Following on the earlier passage with the pocketbook, Defoe again uses plain language in order to force the question of how a social phenomenon might be visible through a natural philosophical lens—or, here, through its basic mode of representation. Seeing just "what happens" is rather derailed in the passage. The most that we can say, ultimately, are the movements and what they are like: racing, dancing, saluting. These mitigate the violence of punishment, even as they lift up its embodied sociality—and the publicity of it. From racing to dancing, Defoe probes actions that are both volitional and controlled from the outside, through social rules or habitual, imitative actions. At most, the conclusion that seems to be offered to us is that we are looking at an action of this kind.

As in the example of the pocketbook, Defoe presents a troubled or inadequate form of perception to leave us with a sense of what is hard to grasp: the social aspect of this set of persons and actions. Crime and punishment—socially conceived actions—are particularly useful for this purpose. In the case of public punishment, however, Defoe deepens his investigation into meaning, calling out a relationship between the social seeing required to grasp the action and another kind of meaning that he presents in the very next paragraph. For even as he leaves us with a mismatch between direct observation and social meaning, Defoe shows us "what that meant" to the two boys. Immediately after the end of the description, the length of the street, Defoe observes, "This was a dark prospect to my Captain, and put him in

Mind, not only of what he was to expect, if he made a slip in the way of his Profession, in this Place; but also of what he had suffer'd, when he was but a Boy; at the famous place, call'd Bridewell." For Captain Jack the meaning is plain: terror at future punishment and a remembered suffering. Only at this point does Jack our narrator call the scene what it is, using terms like "Execution," "Crime," and "Punishment" (all in the following paragraph).

Punishment does not appear to be a focus of the novel: Captain Jack does hang, but we see nothing of it, and Colonel Jack, our hero, escapes unscathed. But Defoe is tremendously interested in how punishment works, both on the criminal and on the larger audience. Such an interest is broad in the first decades of the eighteenth century, even though full-fledged debates over public punishment would not occur until the last half of the century.[61] Of course educational treatises were full of accounts of punishment, and Locke devotes a lengthy discussion to the cons of what he calls education's "last resort." Corporal punishment for the student, Locke says, can backfire: "Passionate words or blows from the Tutor fill the Child's Mind with Terror and Affrightment, which immediately takes it wholly up, and leaves no room for other Impressions" (*E* 222). As Locke indicates, this terror might take the place of other impressions, like the information the tutor wishes to convey. Or it might take the place of the very reform at which punishment, in a domestic case, supposedly aims. We see the problem play out early in the novel, when Colonel Jack witnesses punishment up close in the whipping of the Captain at Bridewell:

> The very Day that we went, he was call'd out to be Corrected, *as they call'd it*, according to his Sentence, and as it was order'd to be done soundly, so indeed they were true to the Sentence, for the Alderman, who was the President of *Bridewell*, and who I think they call'd Sir *William Turner*, held preaching to him about how young he was, and what pitty it was such a Youth should come to be hang'd, and a great deal more, how he should take warning by it, and how wicked a thing it was that they should steal away poor innocent Children, and the like; and all this while the Man with a blue Badge on, lash'd him most unmercifully, for he was not to leave off till Sir *William* knock'd with a little Hammer on the Table.
>
> The poor Captain stamp'd, and danc'd, and roar'd out like a mad Boy; and I must confess, I was frighted almost to Death; for tho' I could not

come near enough, being but a poor Boy, to see how he was handled, yet I saw him afterwards, with his Back all wheal'd with the Lashes, and in several Places bloody, and I thought I should have died with the Sight of it; but I grew better acquainted with those Things afterwards. (*CJ* 69–70)

With terms close to enough to make this a Lockean educational experiment, the Alderman gives the boy moral instruction, even as another man whips him. As Locke predicts, the beating's violence stays with the boy, but the words cautioning him against crime do not. As happens with Locke's student, the punishment "has turn'd his Brains, so that he scarce knew what was said by or to him" (*E* 222). The Captain, at first "sick of the trade" of kidnapping (the crime for which he is beaten), falls in eventually with his old "trade." If any impressions are made, they seem, at least in the Captain's case, to "wear off" (a Lockean phrase popular with Defoe).

In addition to thinking about children and education, Defoe also quietly enters debates over public punishment in this scene. He shows us how the punishment of one boy affects another, precisely the concern of advocates and detractors of exemplary punishment. The notorious Bloody Code, in leveling extreme punishments for small crimes, was understood to function in this way.[62] In the scene at hand, Colonel Jack notes that he was "frighted almost to Death" and "thought I should have died with the Sight of it." He and the Major, he reports, had "sensible Impressions made upon us, for some time . . . and it might be very well said we were corrected as well as he" (*CJ* 70). So far so good for public punishment: reaching through the body of the guilty boy, it has touched young offenders who might have been tempted to commit more serious crimes. But Defoe qualifies this success right away: we are told that within the year the Major has begun pickpocketing again, and Colonel Jack follows shortly.

Impressions, however, are not a good way to track the influence of this particular scene of punishment. We have already seen the whipping at Bridewell return in the much later scene in Scotland, when Captain Jack sees the whipping there. Sometimes the Captain's punishment turns up for Colonel Jack too, as when he hears that a "Companion" has been sent to Bridewell: "then my poor Brother Captain *Jack's* Case came into my Head, and that I should be Whip'd there as cruelly as he was" (*CJ* 84–85). But the scene of whipping operates in less predictable ways than this, and quickly deviates from responses that we might associate with individual psychological reflection. So, for instance, after Jack reflects on Bridewell, it just so happens that, in the very next scene, discussing his theft with his companion Will and becoming distraught over the return of the bills the two have stolen, it is Will who responds, "what would you have me be found out and sent to Bridewell, and be Whip'd as your Brother Captain *Jack* was" (*CJ* 86). Likewise, when the boys are in the square in Edinburgh seeing the public scene of whipping before them, and the Captain reflects on his own condition, a bystander seems oddly to intuit Jack's web of connections, telling the boys that the two men being whipped "were two *English-men*, and that they were Whip'd so for Picking-pockets" (*CJ* 149). In both cases, reflection seems to bring forth associations that exceed the individual's psychology, beginning in a character's thoughts but then circling back through a different character's words.

For the reader the whipped boy of Bridewell—and the "mad Boy" Jack becomes—can seem to be everywhere in the first section of the novel. When Jack fears he has lost the money he has just stolen, he is described as "crying, and roaring like a little Boy that had been whip'd" (*CJ* 82). Then Defoe turns around and uses the same allusion to describe the opposite reaction; when Will finds that there is gold in a bag he has snatched, he "began to Crow and Hallow like a mad Boy" (*CJ* 98). Finally, when the Captain and Jack are tricked into passage to Virginia where they will be sold to a plantation, the Captain, we are told, "rav'd like a mad Man" when he finds out about their capture (*CJ* 160). It is hard to make sense of some of these moments in relation to the original scene of whipping: Defoe seems to draw through an image or idea here, sometimes almost without determined content. But this has its own effect on the reader, who feels that the "mad Boy" and the scene of whipping are somehow everywhere, tying together moments of violence and even exceeding them, filtering into the everyday.

While the mentions of the whipped boy diminish in frequency once the boys leave Scotland, they do not disappear. John Richetti notes that the criminal part of the novel (roughly one-third its whole) "sets the pattern for subsequent adventures."[63] We see this at work even on the level of references to the whipped boy, which link the early criminal scenes with much later ones. For example, when Jack operates as overseer on the plantation in Virginia, he is asked to use a whip but finds it hard to imagine that he "should lift up my Hand to the Cruel Work, which was my Terror but the Day before"

(*CJ* 174). "How," Jack continues, "cou'd I use this Terrible Weapon on the naked Flesh of my Fellow Servants, as well as Fellow Creatures?" (*CJ* 179). Jack's answer is to move away from physical violence, devising a more humane form of punishment that threatens slaves with physical punishment, only to have the master pardon them.[64] But this psychological form of punishment has results that sound familiar. After he lets his first captive, Mouchat, go, the enslaved man comes to Jack "and Sob'd, and Cry'd, like a Child that had been Corrected" (*CJ* 184).[65] Perhaps Jack was able to leap so easily from "slave" to "overseer," from the recipient of the lash to the holder of the whip, because he is already accustomed to such violence, violence that shows up in Defoe's language even when his character claims to avoid it. After all, the more mature Jack, narrating that early scene of whipping at Bridewell, remarks to the reader that "I grew better acquainted with those Things afterwards" (*CJ* 70).

So well acquainted does Jack become that he creates "terror" for the slaves and contemplates—then, enacts—physical violence on individuals once he returns to Europe. Gone for good is Jack's "instinctive revulsion at cruel and excessive violence," which defines him early in the text.[66] When a gentleman comes to collect a bill from Jack's former wife, Jack begins calmly but ultimately confesses that "I had certainly stamp'd him to Death" (*CJ* 243). After a brief stint in the army, Jack reflects on the incident, even more determined on violence: "I should naked and unarm'd as I was have flown in the Face of him, and trampl'd him under my Feet" (*CJ* 248). Later, in Paris, contemplating the cuckoldry of his "*Italian*" wife, he "committed Murther more than once, or indeed than a hundred times, in my imagination" (*CJ* 270, 264). Finally, suspecting that she is in love with a marquis, he "thrust her away with such force . . . [that] she was very much hurt" (*CJ* 269). When Jack deals with infidelity in yet another wife (this one addicted to drink), he finds himself confronting the wife's lover. Here, too, the violence is bodily: he beats the man, only to hear in his cries the Captain's cries, for the man "roar'd out like a Boy soundly whipt" (*CJ* 281). The whipped boy simile travels all the way to Virginia and back. These mentions of whipping link the original punishment of the Captain not only to later scenes of criminal punishment, but to those of slave discipline and domestic violence.[67]

Punishment might prevent crime—or, more likely, delay it (we learn)—but it has a longer and stranger life than this aim for deterrence might predict, one that ties together harsh punishment in England, punishment on the slave plantation, and interpersonal violence. Defoe makes these connections without narrating them, simply dropping in the simile and drawing our minds back, and back again, to the scene at Bridewell. In so doing, he shows us a kind of cycle—a global cycle—of violence. The whipped boy quickly ceases to operate strictly as a point of reflection for the young criminals and instead, moving beyond individual consciousness, shows a world speaking back that violence to him until, ultimately, Jack himself becomes the perpetrator of violence comprehensible in the very same terms. If we return to the passage in Scotland, we see Defoe connect the perception of a social act of punishment (in front of the boys) with a different kind of embodied knowledge particular to the criminal's experience of punishment. The latter, he suggests, is a form of social knowledge, too.

In *Colonel Jack,* crimes and punishments constitute events that allow Defoe to speculate about where we might locate a knowledge of society that exceeds the individual and that could operate on a level beyond his consciousness. In both cases Defoe uses the plain style to show us not just what happens but the experience for which its plainness cannot entirely account. This other experience, outside the ease of plain representation, is repeatedly likened to craft knowledge, whether that of the maker or of the made, a knowledge that exceeds individual consciousness, both in its repetitions and habit and in its collective creation. The artisan-tradesman-thief offers a chance for Defoe to try to reckon with this embodied knowledge, both in its potential—its grasp of the society in which it intervenes—and its horrors, as it registers the effects of society on body and mind.

TWO

Labor

CRAFT POETRY IN DUCK AND COLLIER

DEFOE REVEALS HIS HANDICRAFT PHILOSOPHY by taking stock of the representational practices of natural philosophy and working with and around them to describe the social world. This chapter discusses how the laboring poets Stephen Duck and Mary Collier attempt related work with another genre connected to early experimental science: the georgic poem. When these eighteenth-century poets encounter the georgic, they see both an esteemed classical model set forth by Virgil and an example of a how-to genre in touch with the distribution of natural philosophic knowledge. Virgil's own poem was read—and argued with—over the course of the eighteenth century in practical, agricultural terms.[1] At the same time, the old position of the georgic poet, who works by loose analogy to the laborer, was renewed by Bacon's "Georgics of the mind" and by a novel concept of intellectual labor formed in the context of natural philosophy.[2]

The leap from classical poem to agricultural tract was less extraordinary in the eighteenth century than it would be today. We see this in Chambers's *Cyclopaedia*, where knowledge is divided between "natural and scientifical" and "artificial and technical." Both agriculture and poetry fall under "artificial and technical."[3] Take, too, Samuel Johnson's definition of "art" as "a science" and as "a trade." Here, Johnson at midcentury still works with the definition of art as a translation of the Latin *ars*, and thus pertaining to "professional, artistic, or technical skill, craftsmanship."[4] The Baconian legacy in the print market of eighteenth-century knowledge made popular this meaning of art as the "application of technical skill," with titles covering everything from

The Art and the Pleasures of Hare-Hunting, to *The Art and Mystery of Vintners and Wine-Coopers*, to *The Art of Taking Down Sermons, Trials, Speeches, &c. Verbatim Without Pen and Ink*.[5]

In 1702 Edward Bysshe published his *Art of English Poetry*, its title a translation of Horace's *Ars poetica* and its text a Baconian approach for the would-be poet.[6] Bysshe holds up the "Genius and Judgement" of the poet that, he acknowledges, are "not . . . mine to give."[7] Indeed, in a move supportive of A. Dwight Culler's account of this as the "first handbook intended for the serious poet," Bysshe insists that even the poet of genius will need what he describes as "the Mechanick Tools of a Poet": a dictionary of quotations arranged by theme, a "Dictionary of Rhymes" (the first serious example of its kind) and miniessays on the "structure of English Verses"—clarifying "double Rhyme" and "single Rhyme," for instance, and the role of syllable and accent in the poetic line.[8] Insisting that we not confuse his set of instructions with the desire to make everyone into a poet, Bysshe observes that "a Man would justly deserve a higher Esteem in the World, by being a good Mason or Shoe-Maker, or by excelling in any other Art that his Talent inclines him to, and that is useful to mankind, than by being an indifferent or second-rate poet."[9] Even as he claims that poets must be exceptional, then, Bysshe reminds us that the "Art of Poetry," is an art analogous to many others, like masonry and shoemaking: skills acquired "especially as the result of knowledge or practice."[10]

In this chapter I focus on two poets, Stephen Duck and Mary Collier, who revisit the georgic poem from the position of the artisan, and who consider threshing grain and washing clothes, among other forms of agricultural and household work, alongside the art of poetry. Stephen Duck, the so-called "thresher poet," was surely one of Bysshe's more interesting eighteenth-century readers.[11] We cannot know if Duck breathed a sigh of relief when he read about "the Mechanick Tools of a Poet," or how he thought about the self-instruction he might add with Bysshe's tools to the "natural genius" that the Oxford Professor of Poetry Joseph Spence visited him in Wiltshire to study.[12] But the most famous laboring poet of the eighteenth century could not fail to consider his own physical labor alongside what Clifford Siskin calls the "work of writing."[13] When Duck rose to cultural prominence in the 1730s, his fame spawned a movement. As William J. Christmas puts it, although there had been laboring poets before him, "something coalesced culturally . . . because of" Duck, and as a result "a distinct laboring-class poetic tradition emerged as a significant feature of eighteenth-century literary culture."[14] Among Duck's immediate followers were John Bancks, a weaver's apprentice; Robert Dodsley, a footman; John Frizzle, a miller; Peter Aram, a gardener; Robert Tasterstal, a bricklayer; and, most famously, Mary Collier, a washerwoman.[15] The readers of this laboring-class poetry no doubt were fascinated by a number of different things brought together in the idea of "thresher poet," chief among them the dramatic class rise possible from writing alone and the proof of natural genius by its discovery in unlikely places (the cornfield or the servant's quarters). They were also fascinated, it would seem, by these individuals' attempts to toggle back and forth between different kinds of "labour": the very physical labor of the field or the kitchen and the intellectual labor of reading and writing.

We can see a keen interest in this last phenomenon in Spence's writing on Duck, which introduces the poet to the public.[16] Part of Spence's aim is to account for Duck's acquisition of knowledge under the conditions of laboring: "He work'd all Day for his Master, and after the Labour of the Day, set to his Books at Night."[17] But Spence is intrigued too by Duck's attempt to work harder during the day, thus spelling himself for a few minutes to read the *Spectator* papers. Here, Spence acknowledges relating a detail that might be thought "too particular" by his reader: when Duck stops working and starts to read, "he used to set down all over Sweat and Heat; and has several times caught Colds by it."[18] There is something unseemly about reading while sweaty, but it is also compelling, this visceral account of how little the life of the mind fits with—indeed, how much it may endanger—the life of the body. We might consider Jonathan Swift's famous rebuke of Duck to bring to the fore a similar kind of puzzle, when Swift marks out the thresher's change of occupations for mockery: "From *threshing* Corn, he turns to *thresh* his Brains."[19] Here the humor comes from Duck's mental labor remaining physical, the focus again on the utter incompatibility of the two kinds of work and the danger—here a violent absurdity—of switching between them.

The georgic has played a central role in accounts of intellectual labor in the seventeenth and eighteenth centuries. Clifford Siskin associates the origin of the concept of intellectual labor with literature and with Dryden's translation of Virgil's *Georgics* in particular. As Siskin puts it, the georgic poem "was the means by which the work of writing itself came to be seen as a potentially heroic activity" and indeed a "superior kind of work."[20] While for Siskin writing's superiority is realized in Romanticism, Joanna Picciotto gives intellectual labor an earlier starting point, one that informs Dryden's georgic poem: in the Royal Society's account of the disembodied labor of observation, based on a purified Adamic body.[21] As Picciotto's account stresses, intellectual labor is formed over and against the physical body of the laborer, as in Swift's insult, where the "thresher poet" has his intellectual labor reduced to physical labor, thinking reduced to threshing. Duck cannot escape the physicality that both grounds and is dispensed of by the privileged intellectual sphere that attempts to replace it. This quandary for the laboring poet is also a starting point. Most laboring poets do not have biographers of the likes of Spence, but they are well attuned to the problem of intellectual labor for the very reason that their knowledge projects are perpetually denigrated due to their physically laboring bodies. Indeed, who better to put pressure on an emerging concept of disembodied labor than a man who daringly moved between the physical and the contemplative within minutes, on the margins of a field—and who wrote a georgic poem from the position of the laborer.

The two best-known laboring poets from the eighteenth-century tradition, Duck and Collier marshal the position of the artisan to engage theory and practice through the georgic mode. Craft knowledge, both grounding the new science and simultaneously expunged from it, serves as an important tool for both poets. Their thoroughgoing examinations of the relations between physical and intellectual labor produce more than just a knowledge that the two can coexist: they allow both poets to explore the relation between craft and work, between the body and society. Intellectual labor in this period is usually—think of Bacon or of Sprat—grounded in the craftsman as a male figure, associated with the craft trades and trade secrets. Collier shows us how women's labor embraces craft and calls into question the character of the body that does the knowing, producing a theory of embodied identity as grounded in difference produced by the social world.

Whose Georgic?

Considered in the period an elevated mode of poetic expression, the georgic mode had been newly revitalized when Duck took up his pen in the late 1720s. This was due in large part to John Dryden's enormously popular translation of Virgil's poem, published in 1697, which would see thirteen editions over the course of the eighteenth century.[22] When Duck accessed Dryden's translation in the tiny collection of books on the farm in Wiltshire, then, he would have had in his hands a bestseller whose name means literally "a poem about farming." As John Chalker puts it, in Virgil's poem "the work of the farmer is seen both as the embodiment of permanence and as a foundation of peace and prosperity."[23]

As a georgic *The Thresher's Labour* both participates in and is at odds with the mode.[24] On the one hand, the poem borrows the agricultural setting from Virgil, and it stresses "ingenuity, effort, vigilance" and "experience," making good on many of the themes of the georgic poem.[25] The georgic delivers, as Joseph Addison puts it, the "Rules of Practice."[26] And we see Duck follow the interest in practical knowledge that made georgic poetry "the literary sign of the new science."[27] On the other hand, the poem leaves aside a critical element of Virgil's georgic: the advice to the farmer.[28] Centering the worker, rather than the farmer, is a reversal or "counter-movement," calling attention to the georgic's usual lack of attention to the laborer.[29] As John Barrell has written, most representations of agricultural laborers in the early century, in both poetry and painting, render those laborers in ways that obscure the difficulties of labor.[30] For Duck, the laborer's work moves to the title, and his voice drives the poem.[31]

Duck's contributions—for and against the traditional mode—come together in tackling the georgic's entanglement with natural philosophy from the standpoint of the laborer. As Kevis Goodman has explained, natural philosophy is an easy match with the georgic because of the way poet and laborer are positioned in relation to each other: in Virgilian georgic the poet always works in relation to the ploughman, with Virgil's "versus" pointing us to the relation between the lines of the poem and the furrows in the field. More fully, "*res* and *verba*, things and words, the materials of the husbandman's and the poet's labors respectively, exist at once in a collaborative and a competitive relation to each other."[32] The georgic—in its command to raise the low, to "toss[] the Dung about with an air of gracefulness," as Addison puts it[33]—has always been interested in the problem that the new science brings to the fore: "the power and interest of the real depends on a self-conscious heightening and restatement of the real."[34] This attention, this "virtual reality," allows us to see the poem as a kind of instrument, like a microscope, that renders new the everyday.[35]

Even in Sprat's elevation of georgic husbandry to describe Royal Society philosophy, however, the work of body and hand remains strikingly visible. "It is in *Philosophy*, as in *Husbandry*," Sprat maintains, imagining the group of philosophers who "measure out, and fill into sacks, that Corn, which requires very many more laborers, to sow, and reap, and bind, and bring it into the Barn."[36] Even Addison, committed to shoring up the georgic's identification through high language and "grace," cannot resist a craft metaphor to specify the construction of the poem. When the georgic places its precepts into verse, he explains, "They shou'd all be so finely wrought together into the same Piece, that no course Seam may discover where they joyn; as in a Curious Brede of Needle-work, one colour falls away, by such just degrees, and another rises so insensibly that we see the variety, without being able to distinguish the total vanishing of the one from the first appearance of the other."[37] Just beneath the surface, even the most refined georgic is handiwork.

Critics have noticed, of course, that Duck's poem does not try for Addison's high language and instead breaches decorum with its numerous depictions of "sweat" and its insistence that low forms of work be privileged, a move sometimes described in terms of "realism."[38] But it is worth thinking more about how Duck's decision to write in relation to Virgil confounds the delicate balance of the analogy Virgil's poem instantiates. The georgic depends in its classical mode on the poet's writing along with the laborer but differently: writing poetry and ploughing fields are never equated as kinds of work, never done by the same hands. This is why the georgic works seamlessly with the program Joanna Picciotto describes for the Royal Society natural philosopher, where the combination of "eyes and hands" is realized through a disembodied intellectual laborer.[39] Physical labor structures intellectual labor but only so the latter can achieve its superiority (as theory, as philosophy, even as professional work) through its separation from the physical body. In this natural philosophical georgic, then, to write as the laborer is to disturb the very grounds on which this hierarchy is established and indeed to render suspect the assumption Dryden makes in calling the text of the *Georgics* translation his "Labours," when he dedicates the poem to the Earl of Chesterfield.[40] Moreover, to the purified Adamic body at the core of Picciotto's analysis, Duck and Collier answer with the fallenness of sweat: the Bible's "sweat of thy face" and Milton's "sweatie Reaper."[41]

The Aesthetics of Work

Duck's poem begins with a revelation of the true scene of labor: "Soon as the Harvest hath laid bare the Plains, / And Barns well fill'd reward the Farmer's Pains . . ." (*TL* ll. 13–14). If we believe we are reading a georgic poem, this would appear to close the cycle of labor and balance the "Pains" with "reward," true to the balance of the couplet lines themselves. This is what Dryden's Virgil says will happen: "That Crop rewards the greedy Peasant's Pains."[42] This sounds like a conclusion in Virgil, but Duck's poem not only puts the farmer in the spotlight; it places a semicolon after this reward and continues:

> What Corn each Sheaf will yield, intent to hear,
> And guess from thence the Profits of the Year;
> Or else impending Ruin to prevent,
> By paying, timely, threat'ning Landlord's Rent, (*TL* ll. 15–18)

In these lines the "farmer" reveals himself to be in some control of the full barn but not at all in control of the agrarian capitalism that dominates the English countryside by the 1730s.[43] The farmer is a tenant farmer whose barns are full of corn—or not, he worries—and who is responsible for these profits to the landlord. As Barrell puts it, "the tenant-farmer . . . may well feel himself to be a producer, in relation to his landlord, and poor in comparison with him, while to his labourers he will often appear as the rich consumer of the fruits of their labour."[44] The "farmer," who a few lines later will be called "master," has his own master. This is a world, then, where the corn's "yield" is divorced from the "the thresher's labour" and is thought immediately in terms of "Profit." Duck presents this to the reader through suspending what we believe at first to be a fulfilling conclusion, only to reveal the further actors beyond the farmer on his field. Such reflection makes clear why the poem has been such a favorite with Marxist critics, whose analysis sometimes allows Duck to speak directly to the contemporary historical situation they analyze.[45] Duck, asking us to reflect on labor, the labor that is "beneath the surface" of much art of the period, seems to associate his own project with what we would now call a negative hermeneutics.[46]

There is no doubt that Duck's poem is meant to train us in just this kind of suspicious, critical reading. But Barrell and others have assumed that, as in these first lines, Duck's strategy largely revolves around the visual. Indeed, we recall that Kevis Goodman's account of the georgic links it specifically to Royal Society accounts of vision, which she uses to analyze James Thomson's "philosophic eye" in *The Seasons*. Duck, a reader of Thomson, however, has something very different from this "spectatorial model of intellectual labor" in mind.[47] Roughly two-thirds of the way through the poem Duck offers us the speaker's most extensive moment of reflection on the landscape. On arrival in the wheat field, the speaker instructs his fellow laborers, "Ye Reapers, cast your Eyes around the Field, / And view the Scene its different Beauties yield" (*TL* ll. 224–25). Corey E. Andrews reads these lines to show the "pleasure" that goes along with the "pain" of threshing work.[48] And, certainly, this is a momentary experience of beauty, a view of the field before workers begin the harvest. But the language of the passage signals something else, as well: an engagement with the landscape in terms of the act of perception. Notice that the speaker tells his fellow workers not just to look at the field but to "view the Scene"—framing it, even rendering it artificial, for their aesthetic appreciation of its "Beauties."

To have a reaper instruct other reapers to view a scene is no light matter in the landscape poetry that Duck's lines recall. In a chapter entitled "Being is Perceiving," Barrell, reading both landscape paintings and poetry of the period, associates the leisured aristocratic viewer with what he calls the "high viewpoint from which is visible a deep and panoramic view of a considerable tract of land."[49] In this context the "Beauties" reveal their origin in a disinterested, gentlemanly perspective whose ability to appreciate the beauty of the land derives from the condition of not working it. By contrast, "the practitioners of any particular occupation were assumed to be concerned solely or largely with the immediate ends of that occupation."[50] Duck describes a vision under the sign of pastoral, in which nature (not workers) produces abundance and beauty.[51] Such logic is in play for the aristocrat whose leisure is "a life unproblematically supported by the abundance of nature."[52] For George Lyttelton, Barrell's poetic exemplar of such abundant natural making, "The vale beneath a pleasant prospect yields, / Of verdant meads and cultivated fields."[53] Duck's formulation is similar, as the speaker tells the reapers to "view the Scene its different Beauties yield."

This self-yielding landscape is given to the workers about to make it yield, and at first this seems to offer an aesthetic perspective that could function regardless of class.⁵⁴ Goodridge observes that the lines show the "aesthetic impulse" of the worker, not just of the poet.⁵⁵ "Look," the head reaper seems to say, and you can claim the aesthetic experience that is usually thought to demand that you own this land. This could even be the shift that Jacques Rancière, evaluating an artisan's aesthetic moment, credits with a "redistribution of the sensible."⁵⁶ But having drawn our attention to the issue, Duck takes a different tack. Indeed, the speaker asks the workers to "look again" and to look differently: "Then look again with a more tender Eye," the speaker continues in the next line, "To think how soon it must in Ruin lie" (*TL* ll. 226–27). "Tender" suggests an eye that is sensitive and receptive, belying what happens next, the conception of the field "in Ruin."⁵⁷ When the speaker asks the reapers to look again, the poet harnesses the energy of the first metaphorical use of the eye and turns it back on the landscape with a kind of violence that decimates the field. If one sight elicits productive beauty, the other lays waste, causing "Ruin" and "Desolation." There is more, for Duck follows this passage with an epic simile, ostensibly to fill out further our ideas of the destruction that the harvest will cause:

> Thus, when *Arabia*'s Sons, in hopes of Prey,
> To some more fertile Country take their way;
> How beauteous all things in the Morn appear,
> There Villages, and pleasing Cots are here;
> So many pleasing Objects meet the Sight,
> The ravish'd Eye could willing gaze 'till Night:
> But long e'er then, where-e'er their Troops have past,
> Those pleasant Prospects lie a gloomy Waste. (*TL* ll. 232–39)

The simile describes how we are to view the reapers' destruction of the field: as akin to the pillaging of an invading army. But this is only part of the story told, because within the simile Duck attends not just to action but to perception. With a strange doubleness, the metaphorical destruction, which the speaker/reaper encourages as an act of perception, is explained through a simile that itself is about perception as loss. The simile makes central, that is, not the destruction itself but the destruction of "Those pleasant Prospects." We recall Duck's statement, earlier in the poem, of the worker-poet's difficulty with pastoral convention: "No Linets warble, and no Fields look gay," he asserts and, more generally, and more philosophically, "The Eye beholds no pleasant Object here" (*TL* ll. 61, 56). The philosophical language of poetic inspiration returns again in the simile's lines on destruction. In this "beauteous . . . Morn," by contrast, "*So many* pleasing Objects meet the Sight, / The ravish'd Eye could willing gaze 'till Night" (*TL* ll. 236–37; my emphasis). But the eye's ravishment turns out to be a kind of deception. For although the couplet closes on "Night" and thus seems to follow and enclose the span of a day, the next line undercuts this, and the final lines move backward, as it were, to talk about that view as already destroyed: "But long e'er then, where-e'er their Troops have past, / Those pleasant Prospects lie a gloomy Waste" (*TL* ll. 238–39). In what Duck articulates of the reapers, then, and in the underscoring of destructive perception in the simile, we do not find Rancière's artisan who peers from the window (adopting the landowner's gaze). We find, rather, the adoption of vision by the outsider, the working man, attached to the invader whose destruction does not stop at a view but extends to a way of seeing, leveling the privileged viewpoint itself.

This radical destruction of vision fits well with the antipastoral lines of the poem, which begin in critique but unfold to show us more about Duck's positive vision for poetry and labor. The passage indicates a discontent with aspects of the environment of "tedious Labour" that hardly correspond to the pastoral conventions the poet has in front of him (l. 50):

> Can we, like Shepherds, tell a merry Tale?
> The Voice is lost, drown'd by the noisy Flail.
> But we may think—Alas! what pleasing thing
> Here to the Mind can the dull Fancy bring?
> The Eye beholds no pleasant Object here:
> No chearful Sound diverts the list'ning Ear.
> The Shepherd well may tune his Voice to sing,
> Inspir'd by all the Beauties of the Spring:
> No Fountains murmur here, no Lambkins play,
> No Linets warble, and no Fields look gay;
> 'Tis all a dull and melancholy Scene,
> Fit only to provoke the Muses Spleen. (*TL* ll. 52–63)

With Ralph Cohen, I read these lines as "a series of negations of conventional pastoral themes."[58] Duck lines up conventions and the scene of threshing here, both lamenting the poverty of his own landscape—its lack of even a single "pleasant Object"—and mocking pastoral poetic conventions in the process. The mocked poetry depends on particular kinds of birds and on "lambkins" rather than ordinary sheep. Critics usually read these lines—particularly "The Voice is lost, drown'd by the noisy Flail"—as an indication that Duck wishes to divide physical labor from poetry and, indeed, that he sees the act of physical labor as foreclosing the poetic act.[59]

It's unclear, however, that Duck wishes to write about linnets—even if he could spy one in a Wiltshire field. The speaker laments he cannot "like Shepherds, tell a merry Tale," nor, devoid of "pleasing thing[s]," can the "dull Fancy" bring much to the "Mind." These lines have sometimes been connected to the "realism" debates over pastoral, especially in the famous quarrel between Thomas Tickell and Alexander Pope in the pages of the *Guardian*, where Pope holds up a classical model for pastoral and Tickell shows us what it might mean to modernize the genre.[60] In the very first number of his six-essay discussion Tickell observes that the "Shepherds and Shepherdesses of ancient Times" pursue a "Way of life . . . not now in being."[61] Tickell is clear that pastoral is selective: it should not represent everything and the delusions it offers should be pleasurable ones. But he also, and simultaneously, insists that there must be changes made to the pastoral models as we find them in classical literature: "There are some things of an established nature in Pastoral, which are essential to it, such as a country scene, innocence, simplicity. Others there are of a changeable kind, such as habits, customs, and the like. The difference of the climate is also to be considered, for what is proper in Arcadia, or even in Italy, might be very absurd in a colder country."[62] Such an assumption underlies Tickell's praise for Ambrose Philips's pastorals, with their native flowers, as opposed to Pope's. Still, despite recommending such modifications, such updating, Tickell is clear that there is some core of pastoral that must remain unchanged. For instance, a pastoral poem set at sea—in which the "sea-mews" stand in for the "lark and the linnet"—is a bridge too far for the critic: "yet who can pardon him for his arbitrary change of the sweet manners and pleasing objects of the country, for what in their own nature are uncomfortable and dreadful?"[63]

Duck may appear in agreement with Tickell when he observes that "The Eye beholds no pleasant Object here." Duck agrees, that is, that pastoral conventions are not appropriate everywhere. But the way he states his aversion to pastoral invokes not just the limits of convention but how those limits are expressed through the poet-thresher's mind: "what pleasing thing / Here to the Mind can the dull Fancy bring?" In these lines Duck separates out the poet's "Mind" from "Fancy," the first remaining receptive, while the second is dulled by its surroundings. Now Tickell's account of the pleasures of pastoral includes an account of how close shepherds' minds are to the immediate, physical world. Those "who have little Experience, or cannot abstract, deliver their sentiments in plain descriptions . . . which . . . strike upon the senses."[64] When Duck has his poet-speaker negate pastoral conditions, placing them up against his own observations, he subtly overturns this idea along with the appropriateness of pastoral for his poem. To the idea that the vulgar inhabitant of the scene cannot abstract, Duck lays it on thick, using overtly philosophical language and indeed an abstract conception of "object." Duck uses both "georgic" and "pastoral" expectations to frame his own story about what constitutes experience.

Sweaty Georgic and Poetry as Craft

From sight, Duck turns toward the body as the site of knowledge. In so doing, he starts low. As Bridget Keegan remarks, "No early eighteenth-century poet sweated more, both in and about his poetry, than Stephen Duck."[65] (There are five mentions of sweat in under three hundred lines, as Keegan points out.) Steve Van-Hagen has observed that Duck's lines about sweat "could arguably exist nowhere else in (nonsatirical) verse at this time."[66] Van-Hagen's emphasis here is on decorum, a major issue for georgic as outlined both by Joseph Warton in his edition of Virgil and, more famously, by Joseph Addison who describes Virgil's ability to toss dung with grace. The repeated returns to sweat are a pointed breach of decorum (perhaps one detail too many, as Spence suggests of his own description) and the force of its shock lies both in the word as a departure from the "gorgeous verbal *tekhnē*" of the mode and in the embodiment it assumes.[67] The project of *The Thresher's Labour* is to direct the reader to consider that body as a site of knowledge.

As we have already seen, the poem begins at harvest time and the master's first act is to order the laborers into the barns. Duck then gets right to the work of threshing that we find in the poem's title. Men enter the barn to prepare for the task:

> Divested of our Cloaths, with Flail in Hand,
> At a just Distance, Front to Front we stand;
> And first the Threshall's gently swung, to prove,
> Whether with just Exactness it will move. (*TL* ll. 31–34)

When critics have focused on these lines, they have noticed the men's physical arrangement. Goodridge describes the passage as "dance-like," drawing our attention to the ritualistic quality of this and much other farm work.[68] But Duck's terms exceed those of communal work, stressing additionally the body of the laborer and the precision of its actions. The men's bodies, "Divested of . . . Cloaths," are front and center. Threshing is not understood in the period as an exceptionally skilled form of farm labor like mowing is,[69] but Duck encourages us to think otherwise, stressing the practice and exactitude necessary for the job. Moreover, he insists on this as a kind of practical knowledge. The "just Distance" between bodies is echoed in the description of the "Threshall," swung with "just Exactness." "Just," in both cases, suggests that physical placement or movement: "conforms to a required or agreed standard."[70] What Duck stresses in this repetition, then, is the point at which practice and rule meet.[71] "Just Exactness," moreover, draws us into thinking not about an absolute standard but about process, about how exactness is established through trial and error. If the reader might think of threshing as no more than beating grain with a stick, Duck enables her to think otherwise by breaking down and explaining the science behind the movements. Duck's language of swinging "to prove" sounds experimental, a matter of trial and error, as does his description of up-and-down movements of "equal Force" (*TL* l. 33). Throughout, Duck emphasizes experience as knowledge and the body and its movements as sources of that knowledge.

Duck wraps practical knowledge back into the worker's body, and he adds to this skill and precision the productive valence of the artisan. In Dryden's Virgil, "endless Labour" produces "useful arts." Duck sticks with this idea of productive labor but moves it from the "furrow'd Fields" to a knowledge we might consider in excess of the worker's agricultural production: poetry's rhythms and sounds. First, Duck depicts the sound of the "Threshall." The "knotty Weapons" work this way: "Down one, one up, so well they keep the Time, / The *Cyclops* Hammers could not truer chime" (*TL* ll. 40–41). The Cyclopes are mighty makers, ingenious giants who produce a chariot and a shield in the *Aeneid*. In the example of the Cyclopes' making, Duck stresses the embodied elements of craft. The Cyclopes not only bring together craft and skill; they also bring into the poem bodies and heft. Theirs is a highly embodied kind of craft making, and in this example that work is done by beyond-human-scale bodies, the presence of those bodies felt in the "heavy Strokes" they give with their tools (*TL* l. 42). The Cyclopes are also known for making practical forms of art, as contrivers who fashion things for use.

Sound is made by bodies, and it is felt. The strokes of the "Threshall" and "Hammers" are made material, making "*Aetna* groan"—shaking the volcano itself—and, in the case of the threshers, shaking and moving through the environment: "From the strong Planks our Crab-Tree Staves rebound, / And echoing Barns return the rattling Sound" (*TL* ll. 36–37). The couplet begins with the physical striking of the planks by the wooden tool. But together the two lines move sound between "Planks" and "Barns." Sound itself is "rattling" as the "Planks" themselves must have rattled—the onomatopoeia connecting physical movement and the sound made by that movement. When Duck brings the "chime" of the "*Cyclops* Hammers," that sound recalls poetry, now embodied, associated with physical movement. Indeed, by the time Duck draws out the Cyclopes' "heavy Strokes," he has already laid the groundwork for us to understand "stroke" in several different eighteenth-century senses that bring together music, poetry, and the body. "Stroke" captures the hitting of the hammer on a bell (sometimes in a clock), as well as the physical movement involved in playing a musical instrument. Moreover, "stroke" means a beat, measure, or rhythm (the *OED* quotes "the olde Iambicke stroake" from 1586); it means a movement of the pen (or "brush, chisel, knife, file, etc."); it means a component line of a written character.[72]

As this rhythmic stroke leads us to expect, then, the making of sound is related to the making of time. In Robert Dodsley's "Agriculture," the flails offer "battering strokes."[73] Duck's flails, however, have rhythm and they "chime." The way that the threshers are linked with the Cyclopes through these "blows" that are also "chimes" recalls bells, musical sounds rather than mere thuds, and that in turn enables us to see the time kept as more properly *made*. Moreover, the beats of the "Threshall" are associated with the metrical stresses, the even iambs, in the line of poetry. If the repetition and reversal that occurs in "one, one" reflects the visual up and down, the last part of the line—"so well they keep the Time"—pulls the visual into rhythm, into sound. Threshing is closer to making poetry than the reader focused on those antipastoral lines might have thought.

Threshing also reveals that poetry need not work as natural philosophy does, propping up the intellectual labor of the writer on the body of the laborer. Poetry may well be brought into "drudgery," but this is not damning for the worker-poet or for poetry itself.[74] Duck's knowledgeable, poetic body pushes back both at Royal Society assumptions of the distanced body, as well as those of the georgic poem which, as Goodman has shown us, follows the same set of assumptions. The thresher's revision of the relationship between body and writing also claims for poetry an embodiment that works through repetition, rather than, as in the optics of the new science, the newly revealed world. Duck does all of this from within craft knowledge, reminding us that repetition underlies all craft and indeed is the way in which experience serves to build skillful practice.

Still, this act of poetic making is not fully redemptive. Even as he redefines the site of knowledge within the representation of agricultural work, Duck is well aware that his "master" will not respect this form of making, and indeed this turns out to be an understatement. For the master has no care for process; concerned only with output and price, he is content to take advantage of the threshalls for the purpose of surveillance. If the threshall beats perfect time, then it becomes easy for the master to spy on the workers even though he has left the barn. Indeed, if we use E. P. Thompson's terms, we might say that this situation dramatizes the taking of rhythms of work integral to a task and turning them into clock time, a new kind of measure of labor that Thompson locates in this period. The new idea of time as something you can "waste," the idea that "time is money," is reflected in the co-optation of the beats of the threshall to enforce the continuous work that is now understood to fill time.[75] This new time is explicitly represented, too, in the master's complaint, some lines later, that the workers have "idled half our Time away" (*TL* l. 75). But if Thompson, in hindsight, sees this moment as evidence of the "characteristic irregularity of labour patterns before the coming of large-scale machine-powered industry," Duck helps us to see something else, as well.[76] For he has already encouraged us to think of time in terms of craft, of skilled making. Duck teaches the reader that when the master uses the sound of the threshall against his workers, he deprives them not just of agency or control but of something they have made.

The production of time as craft elevates threshing, drawing it into contact with poetic modes of creation in the making of metrical time. But to think of time in terms of a craft object has further implications for the way we consider labor. For Marx, a century later, craft is the primary way of thinking about labor: "labour process" is "man's activity" which "effects an alteration in the object of labor which was intended from the outset."[77] Although later writers have questioned Marx's reliance on craft as the example of labor—surely, they have argued, labor is much broader and not always materially productive in this way—Marx clearly finds something useful in thinking through this particular form of productivity. Sean Sayers helps us understand why when he points to Marx's reliance on Hegel's understanding of form, which begins very materially—with the plowing of the field—but extends, ultimately, to other kinds of social and political formation. Sayers's own argument, then, is that other readers—particularly Habermas and Benton—take Marx too literally in assuming that the craft metaphor cannot be extrapolated into other forms of labor.[78] The questions surrounding Marx help us to see something about Duck, when he wrests a physical labor metaphor from intellectual labor in order to describe physical labor. For Marx, this material starting point is useful as a way to understand labor's value: the material object provides a means of imagining how labor doesn't just make but somehow inheres in its product. This is the kind of imagining to which Duck leads the reader. Thinking of threshing in terms of a made object, thinking about time itself as a made object, we are led to imagine the way that the strokes of the laborers are contained in objects themselves.

I bring up Marx less to insist on a particular historical connection than to point out that Duck is both describing, something that is often associated with his "realist" mode, and also theorizing, giving us a way to think both about poetry and about labor. I want to return now to the way in which he expands on the idea of poetic, intellectual labor as material. There is another moment in the first half of the poem when Duck explicitly connects threshing not just to making but to writing. After the master has counted the bushels and accused the workers of wasting time, he cries, "Why look ye, Rogues! D'ye think that this will do? / Your Neighbours thresh as much again as you" (*TL* ll. 76–77). Duck expresses the laborers' response this way:

> Now in our Hands we wish our noisy Tools,
> To drown the hated Names of Rogues and Fools;
> But wanting those, we just like School-boys look,
> When th' angry Master views the blotted Book:
> They cry their Ink was faulty, and their Pen;
> We, The Corn threshes bad, 'twas cut too green. (*TL* ll. 78–83)

This is hardly a flattering description of the threshers, men whose "endless" and "tedious" labor Duck has already documented. The comparison to schoolboys seems like one the master might well endorse, both infantilizing and trivializing of what would have been a serious concern for the harvest: the moment at which the grain was cut. At the same time, though, Duck uses this insult to smuggle in an intriguing analogy, one which extends the craft metaphor but turns it toward the concern with intellectual labor. Threshing, Duck wishes us to see, is like the writing in a school book. This analogy goes both ways, lifting up threshing into an intellectual realm—associating it with book knowledge—and reminding us of the very manual ways in which that supposedly higher-level knowledge is acquired: through repetition, through handwork. This last is especially plain in the image of the messy, blotted ink and in the couplet's parallel that attaches the "faulty" ink to the too-green corn: material on which, with which, making happens. Duck reminds us that school knowledge—the very thing that a threshing poet is undoubtedly lacking—is manual, is material, is an art of the hand as well as of the mind. It is hard not to think here of the shoemaker poets Bridget Keegan has followed through the eighteenth century, whose early poems were beaten with an awl into leather.[79]

In the lines following the antipastoral ones, Duck both reinforces this position and also shows us another set of risks or limits to it. The next few lines reveal the poem's most complicated answer to the bodily response that might take the place of this environmentally motivated making of poetic inspiration.

> When sooty Pease we thresh, you scarce can know
> Our native Colour, as from Work we go;
> The Sweat, and Dust, and suffocating Smoke,

> Make us so much like *Ethiopians* look:
> We scare our Wives, when Evening brings us home;
> And frighted Infants think the Bug-bear come. (*TL* ll. 64–69

In describing the dirty work of threshing peas, Duck returns to craft process. The lines dramatize elements from the environment and their transformative effect on the body: "Dust" and "Smoke," combined with the body's own "Sweat," turn the body's color to black. If we think back to Barrell's troubles with the eye, its inability to include the laboring body, the visceral, material response here is striking: not only directing us to the body but insisting on its activity, the transformation effected only when the sweat of labor mixes with dust and smoke to attach their particles to the body.

Duck has us imagine not just this residue of work on the body but the man's return home, looking like an "Ethiopian[]," to scare his wife and child. Given Patricia Akhimie's analysis, we should not be surprised that the White laborer's body and the Black body come together so easily and that they do so on the basis of "somatic markers."[80] In Akhimie's treatment of the early modern period, she describes the "working-class body as racialized, marked as somatically different and understood as naturally unapt for advancement."[81] In the eighteenth century, when these strictures supposedly eased, and indeed in the poem of the man who rose from field to court, we find reiterated the same impossible power relations of that earlier period. Duck in these lines replicates Akhimie's sense of "racialization," pointing to somatic markers as the physical remnants of work. As she points out: "Far from representing inherent baseness, the stigmatized somatic mark is in fact the mark of continual appropriation of economic and cultural capital from disadvantaged groups . . . for the greater benefit of the dominant group."[82]

This exploitative making takes place on the very surface of the laborer's body. Thus, it is both an extension of the craft that has gone before and a warning about the dangers of claiming an embodied identity to begin with. Duck defangs the Black body by showing it a mere "Bug-bear"—imagined fright attributed to a child. But the specter of laboring bodies as bodies that are themselves crafted objects looms, the specter of chattel slavery haunting the poem's domestic argument.[83] Equiano will return to the problem of the enslaved Black body and craft knowledge in this book's final chapter.

Collier and Female Embodiment

Mary Collier includes Duck in the title of her poem, *The Woman's Labour: An Epistle to Mr. Stephen Duck*, and she immediately draws attention to the difference between her position and Duck's. Calling Duck "Immortal Bard," Collier implores him, "Deign to look down on One that's poor and low, / Remembring you yourself was lately so."[84] Collier commands Duck, elevated through the Queen's preference, to remember his laboring identity. She also inaugurates the kind of attack that will dominate the poem. Collier contrasts her own lack of "Learning"—it was not "bestow'd on me," she laments—and her life of "Drudgery" with Duck's success (*WL* ll. 7, 8). Despite the vehemence of such attacks, Collier's poem shows her undertaking a project very much related to Duck's own and to their shared laboring identity, engaging both with the georgic mode and with Duck's experiments in embodied knowledge. There are some aspects of *The Thresher's Labour* that Collier must break open in order to do so, and she leads us through a reading of the earlier poem that uses it to carve out a space for "woman's labour." It is worth stating up front the particular challenge that Collier faces in making her own georgic account, given her employment as a washerwoman. While Sprat and others were keen to use carpentry or husbandry as ways of glorifying the artisan's labor, they are able to do so easily because that craft work is recognized as both skilled and productive. A poem focused on washing, by contrast, poses an interesting challenge for the georgic mode. We do not think of the washerwoman as an artisan; washing doesn't make anything and in fact its work might be understood to be different in kind, as in Hannah Arendt's distinction between *animal laborens* and *animal faber*, between merely laboring and making (techne and poesis) a world.[85]

Collier's poem positions itself as a "response" to Duck's, though we can see quickly the expansive sense in which this poem replies to the earlier one. Most famously, of course, Collier replies to Duck's characterization of female laborers. Whereas Duck depicts women as the "prattling Females" who appear belatedly and who "play" whenever the Master's back is turned, Collier insists that women have been there all along, often engaged in the very same tasks as the men (*WL* ll. 165, 167). To leave no doubt of this, the speaker of the poem turns to her own experience: "For my own Part, I many a *Summer*'s Day / Have spent in throwing, turning, making Hay" (*WL* ll. 49–50).

Here Collier's speaker describes herself engaged in the work of tedding, or drying the hay, reminding us, as Goodridge notes, that it is a skilled task "as important as the mowing."[86] Collier describes, he explains, what we know from history: women and men worked in the fields side by side, engaged in the same or related work. Collier's poetic refutation serves both to quarrel with Duck's image of female laborers and to begin Collier's own georgic poem with the woman participating in the privileged form of agricultural labor from Duck's (and Virgil's) account.[87] Note Collier's term for tedding: "making." Although a "washerwoman" on the title page, Mary Collier begins by establishing women's work in the field and by centering the speaker as a participant who is also a productive maker.

According to Collier, then, women populate these fields and they do skilled physical labor along with the men. The main difference between women's labor and men's labor is not difficulty or required skill: it's that women work harder. In Collier's poem women's labor is always more painful, more difficult, and longer lasting than men's labor, and she makes this argument by attacking Duck's poetic claims to hardship. In such moments Collier is both mocking and competitive, taking Duck's experiences of "endless" work and showing her own to last longer, taking his experience of pain and insisting on her experience as more painful. To Duck's repeated emphasis on sweat, Collier responds: "Not only Sweat, but Blood runs trickling down / Our Wrists and Fingers; still our Work demands / The constant Action of our lab'ring Hands" (*WL* ll. 85–87). Collier takes aim too at Duck's seemingly innocuous lines about returning home so exhausted that he must "walk but slow, and rest at every Stile" (*TL* l. 153). It sounds terrible enough until one comes to Collier, who describes the woman's walk home as including a child under her arm and a domestic world waiting to be tended: "'tis not worth our while / Once to complain, or *rest at ev'ry Stile*" (*WL* ll. 103–4; emphasis in original). Even Duck's account of tortured sleep comes in for this kind of rebuke. In *The Thresher's Labour*, Duck writes,

> Nor, when asleep, are we secure from Pain,
> We then perform our Labours o'er again:
> Our mimic Fancy always restless seems,
> And what we act awake, she acts in Dreams. (*TL* ll. 251–54)

Collier, typically unimpressed, comes back with details about domestic labor and the impossibility of sleep. "Children cry and rave"; "our Mistress sends to let us know / She wants our Help" (*WL* ll. 114, 223–24). In her final couplet of the stanza, she observes, "Our Toil and Labour's daily so extreme / That we have hardly ever *Time to dream*" (*WL* ll. 133–34; emphasis in original). Duck has barely time to rest on the sabbath; Collier has "hardly ever" time to dream even when she does rest. Again and again Collier insists that we hear in Duck's laments a self-indulgent whine.

The witty one-upmanship of Collier's poetry makes a poem about endless labor quite funny at times. It also toys with the concept of experience that grounds Duck's version of the georgic mode. In the lines about not resting at the stile and not being able to dream, Collier lands hard on the difficulty of women's work by reusing Duck's own situations, even quoting his lines. In so doing, Collier asks the reader to register the washerwoman's own reading experience of the lines of *The Thresher's Labour*.[88] *That* is what he calls endless? *That* is what he refers to as pain? Collier makes sure that readers of *The Woman's Labour* are always reading the two poems at once. This readerly experience offers us something distinct from the visibility critics have assumed for Collier's presentation of women's work. Collier does bring women's work to the fore, but she also helps us comprehend the status of this work by making us into close readers of Duck's poetry who grasp not just his bad attitude toward women but the failure of his definition of "labor" to include them. Collier is particularly invested, then, in moments when Duck seeks to represent a universal experience of the laborer that does not speak to her own experience. And she does this by making Duck's account of experience into a kind of norm, something we all know and (as readers of the older poem) take for granted. The example of resting by the stile works in this way. The effect of Collier's response is to reform our original response to Duck's poem and to show us, too, that the laborer's understanding of his own rest is incomplete. As readers of Collier who come to the poem familiar with Duck's poem—a familiarity her title suggests we should have—we feel the gap between the experience related and the one that Collier presents for us, and thus find ourselves in full comprehension of an individual experience that cannot comprehend the experience of others because it cannot accurately grasp its own boundaries.[89]

Collier is clear from the beginning of the poem that this other experience is women's experience, that the speaker's own account captures "The Portion of poor Woman-kind" (*WL* l. 10). By this line, the poem's tenth, we are clear that what follows will use Duck's mistakes to describe a much more general social phenomenon. Collier also includes in the poem's first stanza an answer to why it might be that women receive such a poor "Portion"—from Duck and from the world. That answer involves some strange lines that critics rarely mention, perhaps because they operate in an unexpectedly different register from the rest of the stanza. After Collier has described women's "Portion," she writes,

> Oft have I thought as on my Bed I lay,
> Eas'd from the tiresome Labours of the Day,
> Our first Extraction from a Mass refin'd,
> Could never be for Slavery design'd. (*WL* ll. 11–14)

Women were not designed for slavery, Collier says, or rather "Our first Extraction" was not designed for slavery. This peculiar formulation—where "extraction" seems to refer to the result of the extraction (the "extract")—refutes an origin story that sounds like the one in Genesis. Indeed, five years later Nicholas Robinson would use "extract" in his work *The Christian Philosopher* to describe the biological origin of woman in Eve from Adam. In his twelfth proposition Robinson declares "That *Eve*, the Mother of all Living, was not made of another Lump of Earth, but extracted from the Side of *Adam*."[90]

Robinson draws out a common reading of Genesis: one body was made from the "dust of the ground" and the other, later, from "one of his ribs."[91] It was often assumed from this creation story that women were lesser beings not just because they were "deceived" in Eden but because "Adam was first formed, then Eve."[92] This, at least, is the logic Paul reveals in 1 Timothy. We can see, however, that Collier's lines disturb the clarity of the New Testament verses, both in failing to name Adam's body, referring only to a "Mass," and even calling woman "first Extraction," naming for the secondary act (extraction) its own priority.

This reflection on priority is deepened by the particular eighteenth-century definition of "first extraction" that would have been present for Collier and for many of her readers. For Collier alludes here to the chemical process of extraction, whereby a constituent element is obtained from a substance. Collier's word "Mass" already puts us in mind of this practical scientific context, and it is possible that she would have known the phrase "spirit of the first extraction" from the distillation process involved in brewing beer, one of the household tasks to which she refers in the poem. In any case the term *first extraction* refers to the first fruits of distillation or pressing (beer or wine), and it suggests a very different idea of priority, though an equally craft-centered one, to the divine potter making man from dust.[93] Here, rather than an extraction that removes a seemingly insignificant part of the whole (a rib), chemical extraction would take a piece of the "mass" the same in kind to the whole and would distill a purer element from the smaller mass.

Collier's account of biblical creation is not far from what later readers have claimed as a long-standing misunderstanding of Eve's creation. James Grantham Turner is adamant that in making the claim that Eve's "raw material makes her secondary and derivative," Paul reads Genesis incorrectly: he "uses pre-lapsarian evidence to justify a condition explicitly imposed as a punishment after the fall."[94] What Turner characterizes as a "highly tendentious interpretation" is taken up too by Augustine, who furthers the idea that woman's material secondariness "makes her an ancillary or employee."[95] Turner's disdain for this argument is plain, and he refutes it at the level of the reading of Genesis. After all, he insists in his own biblical exegesis, "Woman, being created from living flesh rather than clay, is presumably a higher being than other creatures—she certainly proceeds from a higher level skill of the artificer, since Adam and the animals were 'formed' and Eve 'built up.'"[96] Collier's lines have a similar project. After all, the "Mass" of Collier's lines is in fact a "Mass refin'd." And it is from this refined material that woman is "extracted," "first" signaling the new kind of creation this could be. Like Collier, Turner keeps thinking in terms of the maker or artificer, and like her, he reads toward secondariness as the more privileged category, here in terms of more advanced, higher-order craft.

Collier, then, responds to Duck's poem that pushes the georgic mode toward embodiment with a revision of woman's first embodiment in Eve. In doing so she lays bare both the social and cultural force of this depiction and suggests the extent to which it underlies all accounts of embodiment, even when those accounts make no mention of bodies that are not male ones. In this Collier does nothing less than remake Duck's account of embodiment from the inside out, building her own account on a revised origin story that is not founded in sexual difference as the reader has known it and that does not reduce the woman's body to this original difference.

Collier places this new account of origins inside a scene of reflection. Immediately before the lines on "Extraction," the speaker observes, "Oft have I thought as on my Bed I lay, / Eas'd from the tiresome Labours of the Day" (*WL* ll. 11–12). For Donna Landry this couplet reveals the speaker's identity as a "poet" who "inscribes herself here as conceiving her arguments for verse in repeated moments of meditation that border on dream-work."[97] If that is the case we must attend to the strangeness of these meditations. Collier tells us that when she does have a moment to lie in bed—something she elsewhere denies that she has at all—her repeated "thought" is that of female ontological difference. We have already seen Duck grapple with the notion that the laborer possesses a body from which mental work cannot be separated (the logic of Swift's account of Duck threshing his brains). When Collier represents herself thinking at this singular moment, she represents reflection in terms of the state of her body (on a bed, exhausted) and including a concern with the material body. Even as Collier proposes a different idea of the gendered body from the Pauline one undergirding Duck's lines, we can see that she also shares the project of moving the body back into intellectual labor.

After she establishes this new account of the material body, Collier re-situates difference, moving it from the creation story's potential for material opposition to a social form of embodiment—a difference, that is, in one's experience of the body. I have already mentioned the way in which Collier draws the women's work in haymaking to the fore, showing tedding as a skilled task alongside mowing.[98] But Collier's inclusions of female labor do not always follow this pattern. Take her account of gleaning, where she adds the female laborer to Duck's male-dominated scene. We find in Collier's new field the marginalized labor of women stooping to pick up the ears of corn missed by the harvesters. Here, however, Collier is concerned that we know more about how this work is accomplished. In completing her work, Collier explains, the female worker must also take care of "Our tender Babes," which means that "round about we gather up the Corn; / And often unto them our Course do bend" (*WL* ll. 94, 96–97). Although gleaning is usually described as taking place with women laborers following their male counterparts down a row of corn, picking up the leavings, Collier's poem characterizes this work as "round about," as deviating from the linear row. She emphasizes the bodiliness of this set of diversions, encouraging us to think about the physical pattern made by the worker in the field, the extent to which the straight path must "bend" to address the child, to create safety in relation to the official laboring task at hand. We imagine the woman leaving the path through a row of corn to chase a wandering child, though it is also clear that taking care of children is "round about" in a different way, perpetually dividing the woman's attention. We can see this in the line "Our Corn we carry, and our Infant too," where the caesura offers us visual balance in the line but the syllepsis belatedly tells us that the "Infant" is also carried, effectively doubling back and borrowing from the verb of the first independent clause (*WL* l. 102). The syllepsis, we might say, forces a "round about" reading, a remaking of verse as "versus."

Children are very present in a poem that is usually considered autobiographical, and we know that it was written by a woman who did not have either a husband or children in her care. Some early feminist readers of Collier were startled that she did not include pregnancy in her definition of "Labour."[99] But as we have seen Collier wishes to establish the materiality of the female body on her own terms, and they are not attached to this biological difference. When she refers to the bending of the women, she references movement and a kind of shared condition. After all, she has given us numerous examples of the ways in which women's work is always "round about" and obstructed. Above all else, this "round aboutness" defines women's labor. Sleep works this way, and women are called away from it by children or by mistresses. The washerwoman cannot enter a house to start her washing because the maid, working late at night, has overslept. Women, as Collier puts it, are "encumber'd thus with Care" (*WL* l. 126).

This is what marks the woman of *The Woman's Labour*: not a different body but a body that is the site of a different kind of experience. In thinking this way, Collier gets close to what Iris Marion Young describes in terms of Maurice Merleau-Ponty's "lived body" and his "conviction . . . that it is the ordinary purposive orientation of the body as a whole toward things and its environment which initially defines the relation of a subject to its world."[100] Young's own interest is in how female bodies make their way through the world, coming up against "opacities and resistances," consistently inhibited intention, precisely because they take themselves up as "fragile."[101] It is not their weak arms but their restrictive social contexts that force girls to perceive the world around them differently and to "throw like a girl."[102] Gayle Salamon, writing in relation to Merleau-Ponty some years later, offers a different though related critique, describing a position of disability as "an embodiment characterized by a kind of dispossession, a loss of that proprioceptive privilege enjoyed or merely taken for granted by those who are not ill or infirm, the ability to unthinkingly and reflexively launch into a project or a world."[103] One of Salamon's conceptions of "this departure, the model of disabled embodiment" has her describe it in terms of a "*bent* arc."[104] Collier's women in the fields with their "round about" movement, are defined by the way they "bend" their "Course."

There are reasons why Collier might land on a reading of work and experience that can be related to much later feminist readings of Merleau-Ponty. As we know from Duck's poem, the eighteenth-century georgic's debt to empiricism draws experience to the fore. Duck himself was critical of the reduction of experience to sight, and we can see Collier's critique as a deeper but related one. We saw Duck's resistance to the mastery of the gaze that sometimes dominates the georgic mode and his attempt to move toward a more bodily mode of perception in order to establish the experience as the laborer's own. It should now be clear why this does not go far enough for Collier. The idea that the "world solicits me," is, says Collier—and after her Young and Salamon—a gendered and ablest assumption.[105] If Collier shares Duck's reservations about a georgic based on mastery, she finds Duck's own resolution too committed to another kind of assumed mastery in its very definition of experience. With this in mind we can revisit the lines on resting at the stile, those lines that show the gap between Collier's lived experience and what she finds in Duck's poem. Collier points out a difference in the body's relation to the world, visible in how a body walks and how it rests.

"Skill and Care"

We should take this felt body into a reading of Collier's account of washing, which the title page declares to be Collier's own work. Collier begins by elevating washing to heroic work, giving it a place in georgic tradition. The washerwomen proceed with "Courage," "Strength and Patience," in the first lines describing their duties (*WL* ll. 156, 158). Moreover, Collier stresses that their work necessitates "Skill and Care," placing it in the category of skilled artisanal labor (*WL* l. 161). Like the threshers, the washerwoman finds her work subject to surveillance, and the women are pressed to complete difficult work in too little time. But there is a critical difference between the two accounts of labor, and it concerns the body. I have already quoted Collier's line that appears in this section, which answers Duck's agricultural laborer's "Sweat" with the washerwoman's sweat and blood. Here are the lines that end that first stanza on washing:

> Not only Sweat, but Blood runs trickling down
> Our Wrists and Fingers; still our Work demands
> The constant Action of our lab'ring Hands. (*WL* ll. 185–87)

This, to be sure, is an intensification of Duck's claim for sweat, yet another testimony to the woman's work as more difficult and as more "constant." This line also reminds us of something about Duck's laborers: in his description, their bodies are never harmed from their labor. The male laborers' bodies remain intact; sweat doesn't disturb the confines of the body. But in washing—and in cleaning and polishing metal pots—the woman's body is injured: "Our tender Hands and Fingers scratch and tear" (*WL* l. 217). Duck's destructive "Eye" was "tender," but Collier claims this delicate sensibility—this ability to feel and to be harmed—for her hands. When Collier calls attention to her sweaty and bloody hands, then, she not only displays her own work as more physically difficult than Duck's; she calls attention to a kind of injury that is strangely impossible for the thresher, given the way that he attempts to reembody labor as a part of the heroic georgic mode.

Collier shows women's experience in moving through the world. But she is invested too in producing a version of the body that highlights abjection, claiming Duck's sweat and adding to it blood. The competitive framework in which Collier approaches Duck extends as well to her account of blackness, which responds directly to the thresher's "Ethiopian" persona who enters the house and frightens the woman and child. To this, Collier writes that polishing pots creates an even more extraordinary situation for the charwoman:

> Colour'd with Dirt and Filth we now appear;
> Your threshing *sooty Peas* will not come near.
> All the Perfections Woman once could boast,
> Are quite obscur'd, and altogether lost. (*WL* ll. 219–22)

The washerwoman who harms her hands in an effort to not "*leave the Dirt behind*" (these are the words of her mistress) now finds herself covered in dirt, much dirtier, much filthier, than the blackened man about whom Duck wrote (*WL* l. 175; emphasis in original). This is not a poem about female "Perfections," but Collier raises female beauty here in order to show the extent of what has been done to the body: "Colour'd" (as in Duck) black, perfections are "obscur'd" then quickly, just after the caesura, wiped out entirely.

In the lines about sweat and blood, and in the lines about dirty blackness, Collier's replies to Duck's poetry take on a different valence. Here, we might say, the goal has exceeded proof of harder work and has entered a zone of competitive abjection. This, I think, accounts for the strange response a reader might have to these lines, where Collier represents a form of greater degradation or pain as in fact a kind of triumph. If Duck aimed to defend himself against the connection with the Ethiopian, Collier presents a very different logic here, embracing and moving forward from this debasement. We are reminded of the potential for abjection, since Kristeva, to be used as a way to recover agency, for Collier uses her overall combative approach to Duck's poetry to emphasize a degree of extremity that seems as though it is debasing or lowering to be a kind of witty triumph, a win against Duck if not against the world.

Collier's account of the female body and work in terms of abjection has an interesting relation to another georgic-inspired poem of the early eighteenth century, one which like Collier's features "Rivulets of Sweat," a blackened female body, and a speaker who cautions not to "Let the Beaux approach too near": Swift's "The Progress of Beauty."[106] Whether or not Collier knew this poem, its engagement with craft labor and the female body further illuminates Collier's own project. Swift's poem is usually understood as an "often savage critique of women's embodiment."[107] The poem features Celia's attempts to "reduce" her face to beauty and ends with the failure of that project in the complete collapse and decay of her body, no longer able to be kept up. As Tita Chico notes, the poem's allusion to syphilis as part of bodily decay also makes clear the extent to which the poem sees beauty as "an effect of aesthetic and social orders."[108]

As Jonathan Kramnick has written, however, we can also understand Celia as an artisan, her cosmetics as materials for making. Kramnick's redemptive reading of Swift's poem draws our attention to Swift's deliberately artisanal terms for Celia's tools and process. When Celia puts on makeup, she uses "the stuff of the world," and she is engaged in "the turning of potential beauty into actual shape, forming we might say rather than form."[109] What she creates she then beholds, "As Other Painters oft adore / The Workmanship of their own Hands."[110] This is "art and craftwork," Kramnick writes, and decay is natural, "once forming reaches its end."[111] Chico, for her part, observes the way that the poem founds social order on the female body's excessive materiality and, indeed, lack of order. If we bring this into Kramnick's reading, we can see a major divide between the two critics, less over artisanal work than over which actor forms the body. What Kramnick assumes is the genuine work of Celia, an individual artisan on the model of the male artist, Chico highlights as the way that the female body is worked on, indeed used as the occasion not for individual but for cultural production.

Collier has a stake in this split over the identity of the individual artisan and indeed about the social register in which the woman's body is to be read. When Collier embraces the abject aspects of bodiliness—blood, filth, blackness—she does not exactly grasp disorder or natural order, either one. She grasps the remnants of order making, the castoff waste of subject formation—this very literally, in comparison to the subject making of Duck's poem. Collier also divorces the charwoman's own labor from this process. Celia helps us to see the force of Collier's claim that the differently "colour'd" female body is made through work. Work discipline is the shaper of the female body; the disorder of Swift's poem would be read by Collier as forced on the charwoman from the outside. But Collier's answer to this is not found in Swift's poem: the worker not only opts not to fix the violation of beauty; she seems intent on undoing the assumptions of value generated in her own account of beauty lost. This can make the claim of losing into a kind of triumph. As the artisan, the skilled laborer, Collier shows us the trouble there might be if we assume women's making is assimilable to the model of the male artisan (in Swift's poem, or in Duck's): it could blind us to the very sorts of injustice with which she and Duck are both concerned.

Above all, washing is Collier's opportunity to redefine what both the georgic and Duck claim good labor is: productive and transformative. We may look too at the way that Collier's labor does not remake the world, in the hope that we may see what it does instead and how that doing is related to poetry. Note how the "Skill and Care" emerge in the account of washing:

> Heaps of fine Linen we before us view,
> Whereon to lay our Strength and Patience too;
> Cambricks and Muslins, which our Ladies wear,
> Laces and Edgings, costly, fine, and rare,
> Which must be wash'd with utmost Skill and Care;
> With Holland Shirts, Ruffles and Fringes too,
> Fashions which our Fore-fathers never knew. (*WL* ll. 157–63)

Poems and novels of the eighteenth century are notoriously filled with objects: not just Pope's "bibles and billet-doux" but those on Robinson Crusoe's lists and in Pamela Andrews's bundles. As Chloe Wigston Smith puts it, "The relationship between the novel and material objects has been considered reciprocal, in which literary and material culture together engaged the expanding commodity culture of the period."[112] The list is a powerful form for doing this. Although Collier's pile of laundry is not quite Belinda's dressing table, full of cosmetics from around the globe, we do feel geographical consolidation in the precision with which Collier makes her way through France ("Cambricks" and probably the lace) India ("Muslins") and Holland ("Shirts"). The "Heaps of fine Linen" gesture toward the abundance of these luxuries, and "costly, fine, and rare" reminds us that they are in no way pedestrian goods.

The construction of the lines of poetry seems to accentuate the objecthood of the dress, the commodities. We see the "Heaps" before their view is attributed to the washers, and the other items display themselves similarly, falling at the beginnings of lines. Even the words on the page, then, bear some resemblance to the "proliferative listing" that Laura Brown describes in *Ends of Empire*, our eyes skipping lines to group the capitalized objects' nouns.[113] And yet in Collier's poem we do not end with these goods, even if poetic form suggests that we should imagine it (for the speaker or for ourselves). Rather, we are led to imagine the cleaning of these garments, the work that is done simply in order to maintain them. The triplet is instructive here. The first line lists "Cambricks and Muslins, which our Ladies wear," and the second follows suit with its similar construction, capitalized words up front followed by a caesura: "Laces and Edgings, costly, fine, and rare." But the third line—and this is the only triplet in the stanzas on washing—does not follow this pattern, turning instead to the laborer: "Which must be washed with utmost Skill and Care." This is skilled work, but it is not making: Collier is not a seamstress or a weaver. Her work, washing, is, rather, maintaining, keeping up. It does not transform the world as the artisan's labor—including that of the thresher, linked to Vulcan's artistic productions—is thought to do.

In Collier's "Skill and Care," she specifies the kind of labor that washing and other charring work entails. The term *cura* or *care* is Virgil's, and its most obvious use is to specify the farmer's care for the land. As Janet Lembke explains in the introduction to her translation of Virgil's poem, georgic care entails "caring without cease for the land and the crops and animals it sustained."[114] The term *cura*, as David Fairer has remarked, goes significantly further than this. He describes the natural world in georgic as needing "to be subdued by work." *Cura* in this project, may "combine caring, managing, and toiling." It is, he says, "both the painstaking and the painful."[115] It also stresses the element of "co-operation," as opposed to "mastery," which makes it an excellent georgic word with which to redefine Duck's (and the eighteenth-century georgic's) project.[116]

In Collier's passage on washing, "care" refers to the gentle precision demanded by the fragile luxury fabrics the washerwoman encounters in her pile. But Collier has used the term two stanzas earlier and repeatedly: it appears four times in the fourth stanza, each time at the end of a line. Here the words refers to the "frugal care" of gleaning corn; the "greatest care" with which children are put to bed; the "care" with which the children are brought to the field; and, finally, it sums up these actions when Collier describes the work of women as "encumber'd thus with Care" (*WL* ll. 100, 109, 120, 126). We notice that in this fourth stanza, Collier does not use "care" in the sense of care for the land. Rather, she moves the term toward domestic tasks and childrearing; it comes to signal both an affective attachment to the child and the management of the household (including the child but also food). Like the careful washing, these actions are also attentive to detail, marked by careful attention, the turn of the woman's deliberate thought and action.

Make no mistake, Collier tells us, care is a burden—a literal encumberment in that line, which draws back into our discussion of the woman's body as hampered in its movements. Care, however, is also an asset, as its prominence in stanza four lets on, and it is something that Collier means to specify in women's work in particular. When Carol Gilligan suggests that women might have a different kind of ethical stance, one focused more on "care" than on "justice," she extends an argument that we can see Collier begin to make here.[117] For when Collier reframes women's work around washing, she reminds us that although craft theory enjoys accentuating the transformative, indeed, the forming process of craft work, that same craft work also requires less spectacular formal investments: repetition, even preservation or maintenance as opposed to making something new. Such work can be exploited like any other laboring act, as we see in the scene of washing clothes. It also exceeds, stands at an angle to, the productive labor of the artisan and the male agricultural laborer. Whereas Duck could not figure out how to disentangle his own way of making, founded on craft work, from the scene of supervised labor, Collier here presents an alternative: linking the care of supervised work to the care that happens around the scene of labor, in its interstices or cutting across its straight lines.

Being careful, in Richard Sennett's account of craft, involves "learning from things," very much an account of the good life that we might find in Virgil's *Georgics*.[118] But Sennett stresses too the extent to which craft is an engagement with the world, one that involves problem solving that can fit us for human relationships: "Both the difficulties and possibilities of making things well apply to making human relationships."[119] Collier's craft poetry heads in this direction when it moves "care" between objects and humans, between gleaning and caring for a child. To mastery, the female laboring poet answers with ethics, with an account of work and a personhood that is embodied differently.

THREE

Art

HANDICRAFT LINES IN HOGARTH

THE "ART" THAT ALLOWED DUCK and Collier to argue for the proximity of writing, threshing, and washing, would not make it through the century. Ephraim Chambers's *Cyclopaedia* of 1728 reflects the capacious reach of "art" on which the laboring poets (following the natural philosophers) drew: knowledge is divided into two major branches, "*Natural* and *Scientifical*" and "*Artificial* and *Technical*." The latter "applied" knowledge contains everything from poetry and painting to agriculture and metallurgy.[1] But by midcentury a split was emerging between the "beaux-arts" or "fine arts" (as they would come to be called in English) and the more practical "useful arts" or "crafts."[2] Larry Shiner argues that we can see a large European shift in this direction beginning around 1750, and it is easy to map Britain's place in this trajectory: the Society for the Encouragement of Arts, Manufactures and Commerce posited one division of knowledge in 1754, to be countered by that of the Royal Academy of Arts in 1768.[3]

William Hogarth, whose aesthetic treatise *The Analysis of Beauty* (1753) is the subject of this chapter, was a central actor in the local London-based struggles over this division of knowledge.[4] The year after the *Analysis* was published, Hogarth was elected to the Society for the Encouragement of Arts, Manufactures and Commerce (hereafter, as in the eighteenth century, the Society of Arts), whose name spells out its economic project, one determined to capitalize on skill and invention to address Britain's weaknesses in global trade.[5] The Society's mission is solidly Baconian in turning "art" (including natural philosophy's productions) toward trade.[6] At the Society of Arts' independent founding in 1754, there were four Royal Society Fellows among its members, and the full group possessed interests and expertise ranging from optics and astronomy to the manufacture of jewelry and porcelain. The Society's first premiums, or monetary awards, reflect the world of knowledge it encompassed. It offered cash awards to practical drawing (the best by a boy or girl between fourteen and seventeen years old), a skill the Society hoped would directly aid in manufacturing.[7] The Society also offered cash awards for the development of the production of madder and cobalt, the first a red dye to be used in textile manufacturing and the second a coloring material for glazes and enamels.[8] The Society would support painters (Hogarth was the second to be elected), but such artists were considered part of a group of "ingenious mechanics" that Henry Baker, a fellow of the Royal Society and founding member of the Society of Arts, spelled out in detail: "carvers, joiners, upholsterer, cabinetmakers, coach-makers, and coach-painters, sign painters, weavers, curious workers in all sorts of metals, smiths, makers of toys, engravers, sculptors, chasers, calico printers, etc. sailors that can take the bearing of coasts or the plans of harbours, and soldiers better qualified for becoming engineers."[9] In considering printing calico alongside surveying coastlines, we certainly feel the force of Chambers's applied sciences put to the singular mission of advancing England's economic and imperial interests.

Hogarth would have been right at home with this combination of practitioners. He had developed his skill in engraving through an apprenticeship to Ellis Gamble, a silver engraver, at sixteen.[10] This unambiguously artisanal training may well have informed his basic conception of the "arts."[11] By the time of his election to the Society, he was a famous painter, well known as the rare painter who could—and would—engrave his own work.[12] Even after achieving recognition as a painter, Hogarth relished locating the origins of his own work in ancient English signboards and graffiti.[13] Jenny Uglow observes that, as he contemplated the *Analysis,* Hogarth visited the public markets for sign painters' work.[14] During these years Hogarth was widely known for his leadership at St. Martin's Lane Academy, which supported medallists, engravers, and enamellers, practitioners of what we would now call the "decorative arts." This was an academy, not a craft workshop based on the apprenticeship model, but Hogarth from the beginning embraced the commercial side of art, his own profitable engravings included. His mentor James Thornhill became famous as a history painter (in some eyes the most distinguished of painterly genres) but then began to work outside those bounds and is now credited with some of the earliest conceptual thinking on interior design.[15]

Hogarth wrote the *Analysis* with debates around the formation of an English academy of arts raging.[16] For some years it had seemed possible that England could finally establish its own academy on the French model. Hogarth had already set himself over and against such a continental academy at St. Martin's Lane, abjuring the traditional hierarchy and emphasis on copying statues and forwarding instead a more democratic setting that encouraged drawing from life. When the Royal Academy of Arts was founded some years later, in 1768, it would include only painting, sculpture, and architecture. At its head, Joshua Reynolds would defend painting as a "liberal art" precisely by separating it from the wrong kind of handwork: "mechanical felicity" in drawing.[17] As John Barrell explains, the distinction between the arts of the Royal Academy and other arts is consistently represented as a difference between "the mind and the hand."[18] We recognize the division as Aristotle's separation of theory from practice, episteme from techne. Like the French academy, the project was not only to narrow the "arts" but also to raise the artist's social status, establishing "'system and order'" and lifting the practitioner "above membership in a craft guild."[19]

Hogarth responds in terms that we recognize from this book's other handicraft philosophers: by turning back to natural philosophy and by taking up the knowledge of the artisan. Hogarth begins his written treatise by telling us that he has hardly picked up a pen and that he writes—over and against "mere men of letters"—as an author with "a practical knowledge of the whole art of painting (sculpture alone not being sufficient)" (*AB* 1, 2). Hogarth begins the treatise by embracing Baconian empiricism and proceeds to lay out beauty in a wide range of objects, both the classical statues privileged by the European academies, and the art objects of interest to the Society of Arts: table legs, smoke jacks, women's stays. In Hogarth's "beauty," the craftsman's creations loom large, and they come with as much potential to teach the reader about beauty as do the *Apollo Belvedere* and the *Farnese Hercules* of the first plate.

In this chapter I track how Hogarth moves his broad definition of art and the work of the artisan into his definition of the beautiful object and into his account of how that object, in turn, is connected to the world around it. This is a theory of the artisan and, from Hogarth's account of painting through to his account of perception, it involves the body and bodily making associated with handicraft. Critics have found Hogarth's theory of beauty—particularly its foundation in a single "line of beauty"—hard to square with the local and cultural emphasis of his paintings and engravings, which are committed to centering the ordinary people and events of eighteenth-century London. Recently Abigail Zitin has offered what might at first seem like a solution: we might raise Hogarth's philosophical profile by completely detaching his treatise from his art.[20] In what follows I argue for the impossibility of quarantining Hogarth's practice from his theory. Moreover, Hogarth's handicraft aesthetics offers us a way to see old epistemological connections between thinking and making as grounds for a formalism that attaches form to the world. As is appropriate for the theoretical work of a declared practitioner, Hogarth formulates this theory in his written treatise as well as in his "explanatory plates," tools of visual explanation that are also the material products of his own artisanal engraving.

Useful Beauty

Hogarth sent a copy of the *Analysis of Beauty* to the Royal Society after its publication,[21] and he makes plain early in the text his investment in natural philosophical accounts of knowledge.[22] In the introduction he tells his readers that, unlike the connoisseurs, we must "*see with our own eyes*" (*AB* 18). In depicting the kind of knowledge he's after, Hogarth turns back to Bacon's address to the reader in the *Novum Organon*. There, Bacon cautions us to turn away from the "artificial *idols* [that] have entered men's minds either from the doctrines and sects of philosophers or from perverse rules of proof," and to turn instead toward "the evidence of our own eyes."[23] Like Bacon, Hogarth wishes to show the problem of received knowledge as a problem of vision. Hogarth's connoisseurs have their idols too, for their "thoughts have been entirely and continually employ'd and incumber'd" with learning of a misleading sort: the "*manners* in which pictures are painted," for instance, or "the histories, names, and characters of the masters" (*AB* 18–19).[24] So taken over are these minds that the viewers' eyes themselves become corrupted: they have "eyes less qualified for our purpose" than the untrained ones of the common reader (*AB* 18). Hogarth wishes to turn all his students, like the artists at St. Martin's Lane, away from classical examples and toward life. As

Diana Donald puts it, "he undoes earlier art treatises' focus on antiquity and points us toward 'nature itself.'"[25]

Seeing with our own untrained eyes is quite a radical proposition within the context of art treatises like those Hogarth names in the preface. Rather than only providing rules to painters, as did Leon Battista Alberti and Giovan Paolo Lomazzo, Hogarth also approaches beauty from the spectator's experience.[26] But if it pushed against the grain of the continental treatises, Hogarth's claim to experience feels a bit hollow—or at least belated—when it comes to the other immediate context for his treatise: the recent British tradition of aesthetic philosophy in the work of Anthony Ashley Cooper, third earl of Shaftesbury, Francis Hutcheson, and Joseph Addison.[27] In that tradition, after all, "experience" of the world is a given as the starting point for aesthetics. As Hogarth marks it out here, such a starting point reorients aesthetics toward the spectator's or viewer's perception, rather than the artist's or the critic's. But it is clear to philosophers right away that "experience," while it can offer an enticingly immediate account of our feeling of pleasure in an object, can be challenging to harness for "beauty." For instance, does one experience beauty as one hears sounds or sees colors? The most notoriously awkward philosophical solution to this quandary comes from Hutcheson, who invents a separate interior sense, one that George Dickie explains is an "affective" sense (as opposed to vision, for example, which is both cognitive and affective).[28] Hutcheson was keen to make his argument for the internal senses by calling up examples of beauty in things like theorems and universal truths, in which the external senses do not obviously have a role.[29] The worries about the relation of beauty to the five senses does not end with Hutcheson, even if his particular solution does not travel far. Later empiricist writers who do not posit an internal sense can still be found to echo Hutcheson's reservations, as when Addison describes the "pleasures of the imagination" as "not so gross as those of Sense, nor so refined as those of the Understanding."[30]

When Hogarth says that we should just open our eyes, then, the reader of Hutcheson might well answer back that things are substantially more complicated. While Hogarth does not refer to any of his immediate British precursors by name in the *Analysis*, he opens his treatise with a question that they had all entertained: what is the relation between beauty and utility? In "Of Fitness," Hogarth approaches a classical idea about beauty that Hutcheson had tackled as part of his account of immediate aesthetic response.

As Robert R. Clewis and Paul Guyer have shown, by the time of the *Analysis*, Hutcheson and George Berkeley had examined beauty's old relationship to utility through the lens of empiricism.[31] Hutcheson, for his part, recognizes that considering an object's utility is impossible in the immediate response that is necessary for his universal beauty. For utility to enter the judgment of beauty, the mind would have to summon a faculty like Locke's understanding, which could then evaluate the example according to criteria for "Use."[32] Berkeley, for his part, argues against Hutcheson that "beauty is . . . an object, not of the eye, but of the mind." Just as he would argue that three dimensions are learned, not merely seen, Berkeley argues that proportions ("as . . . they may best conspire to the use and operation of the whole") are accessed "only by reason by means of sight."[33] Hutcheson comes back hard. One of his strongest examples is that we can see the difference between beauty and use in a chair where "the Feet . . . would be of the same Use, tho' unlike." This ill-designed object might have equally long feet but "one were strait, and the other bended; or one bended outwards, and the other inwards."[34] Hutcheson is certain that this would not be judged a beautiful chair.

Hogarth begins his chapter on fitness (that is, he begins his entire treatise) with a paragraph that seems directed at just such an example.

> Fitness of the parts to the design for which every individual thing is form'd, either by art or by nature, is first to be consider'd, as it is of the greatest consequence to the beauty of the whole. This is so evident, that even the sense of seeing, the great inlet of beauty, is itself so strongly bias'd by it, that if the mind, on account of this kind of value in a form, esteem it beautiful, tho' on all considerations it be not so; the eye grows insensible of its want of beauty, and even begins to be pleas'd, especially after it has been a considerable time acquainted with it. (*AB* 25)

Hutcheson and Berkeley help us to see that Hogarth's "fitness" presents a beauty that must allow the intervention of the spectator's mind to understand the workings of the object and to judge those workings "useful" or "fit." In this passage Hogarth attends to the "sense of seeing," but, like Berkeley, he is occupied with the mind's involvement. Mind speaks back to eye, even forming its "bias." On the condition that mind "esteem it beautiful," recognizing in the object the fitness Hogarth has defined, then the eye will come around. Thumbing his nose at the sort of immediacy Hutcheson desires for his universal beauty, Hogarth tells us the response to fitness could take a while: the eye first "grows insensible" and then "even begins to be pleas'd." If the table is useful enough, that motley combination of legs might begin to grow on us.

In his passage on the mind and eye's response to troubling proportions, Hogarth ends up on Berkeley's side. His account sets Hogarth firmly apart not just from Hutcheson but also from Shaftesbury, whose beauty was established over and against utility. Hogarth sets himself apart too, of course, from Kant's definition of beauty in the *Third Critique*, a common source for theories of beauty ever since (and part of our own surprise that beauty and utility go together in these earlier texts). By sharp contrast with Kant, Hogarth advocates for a beauty that reaches out to include, even privilege, the conceptual.

The risks of this, as Hutcheson knew quite well, are serious. One of the reasons that Hutcheson worked hard to keep out utility has to do with the kind of conceptual ideas one must import to comprehend use. Take Hogarth's own examples of the racehorse and the warhorse, which he modifies from Xenophon's *Memorabilia*.[35] Hogarth alerts us to the fact that there is no single beauty here: the beauty of the racehorse requires one body, that of the warhorse another. Like Xenophon's Socrates, Hogarth asserts that beauty must be a beauty *of something*, directed toward some use. This "beauty for" can at least crack open the door to relativism, as Guyer indicates was a worry for Hutcheson.[36] After all, use points to how things are usually done (one might add, within a particular context at a particular point in time).

For now, let's leave that door cracked open. Hogarth's immediate goal is otherwise: to show us some examples of "fitness" that make clear the complexity of the attachment of a concept to a beautiful thing. In the very short chapter on fitness, Hogarth uses the term to pull together a range of examples of beauty, typical of the great variety that the treatise will exhibit and that we might already recognize from a quick glance at the plates. Hogarth offers us examples of "fitness" (or, as he says repeatedly, "fitness for use") in the following order: architecture (columns); domestic objects ("chairs, tables, and all sorts of utensils and furniture"); architecture (pillars, arches, windows, doors); ship building; the human body; race- and warhorses; and "the Hercules, by Glicon" (*AB* 25–26). As he will do throughout the treatise, Hogarth positions what appear to be different-in-kind objects side by side. He attends to classical columns, then turns to chairs and tables; he has us think about ship building and then he turns to bodies (of humans, of horses) and, finally, to a famous classical sculpture.[37]

In this chapter, as previous readers have noted, Hogarth draws on Xenophon's *Memorabilia*, where Socrates considers the relation between beauty and use. In perhaps the best known exchange in book 3, Socrates discusses a golden shield and a dung basket. Aristippus asks him, "Is a dung basket beautiful then?" Socrates answers, "Of course, and a golden shield is ugly, if the one is well made for its special work and the other badly."[38] Socrates clarifies that beautiful objects are good *for* something: for fever, for running.[39] With Socrates' discussion of the "well made," it makes sense that he would eventually turn to Pistias, the armorer, to discuss utility with an artisan. Hogarth must have been pleased at the way Socrates puts his question to Pistias, for he asks why Pistias charges more money for his breastplates. The answer does not have to do with the beauty of the decoration on the breastplate but with its fitness. Socrates comes to this in two different ways, both of which are important for Hogarth's passage. Socrates starts by noting how "beautiful" the breastplate generally is as an "invention": it "covers the parts that need protection without impeding the use of the hands."[40] But when he asks Pistias about the particular value of his breastplates over others, another account of beauty as fitness emerges. Here, Pistias refers to better "proportions," as will Hogarth in his chapter.[41] After an exchange, Socrates confirms that Pistias's answer means "well proportioned not absolutely but in relation to the wearer."[42] Socrates explains further: "The good fit is less heavy to wear than the misfit, though both weigh the same. For the misfit, hanging entirely from the shoulders or pressing on some other part of the body, proves unwieldy and irritating; but the good fit, with its weight distributed over the collarbone and shoulder blades, the shoulders, chest, back and stomach, may almost be called an accessory rather than an encumbrance."[43] If Socrates begins by thinking about the beautiful and the useful in terms of function, he expands that in his exchange with Pistias. The breastplate has a general function, it is true, in protecting the body and leaving the hands free. But "good fit" here, is also established in relation to the body. Socrates makes clear that such beauty stands over and against the kind of beauty we might find in the ornamentation on the breastplate itself.

Both of the breastplate's forms of fitness come into play in Hogarth's account. Hogarth points in a slightly different direction from Socrates, however, in his attention to the way we perceive utility or fitness in the object. His first example is "twisted columns," which "convey an idea of weakness" and thus "always displease" (*AB* 25). This first example from architecture, then, begins by instructing us that we must be able to discern the purpose of the object in the form it takes. Thus, "twisted columns" may hold up a building just fine, and their shape could be beautiful in the abstract, but beauty requires something more: as Socrates indicated, it must be beautiful in the context of its use. Here, then, Hogarth complicates any notion we have that form could be reducible to mere shape. And for beauty, he recommends that we not go the route of the ugly table. Rather, we should try for a form that communicates the strength required by the object's purpose.

We notice that "fitness" for Hogarth is, like the rest of the treatise, invested in how beauty is perceived. We can see Hogarth move closer to the breastplate in his next set of objects: tables, chairs, and utensils. As with Pistias's breastplates, these items are "fit" not just in themselves but in relation to the human body. So strong is the body's connection to such items, Howard Risatti tells us, that a chair keeps signaling its intended function, even if it is being used for something else.[44] For Risatti, this is an example of how craft objects contain their purposes within them. And it is an example, too, of how craft recalls the body doubly: of the artisan who makes it, and of the use it will come to in the world after it is made.[45] We might note that Hogarth turns architecture in this direction too. He writes first of the "dimensions of pillars, arches, &c. for the support of great weight," but then he turns to "the sizes of windows and doors, &c. Thus though a building were ever so large, the steps of the stairs, the seats in the windows must be continued of their usual heights, or they would lose their beauty with their fitness" (*AB* 26). Again, we see the resemblance between his account of "proportions" not as beautiful in the abstract but as useful or fit to the purpose of the body that will use the object.

When utility gives abstract rules a push, it does so by insisting that we think about the object's purpose in a social world that we inhabit. What other way would there be to think about a dung basket or a shield? Grasping that purpose, as Hogarth has already begun to suggest, and as Socrates did in his visit to Pistias, requires a form of maker's knowledge. For the object is not made useful merely in its shape but also in its materials, in the finer points of its making. Risatti in his much more recent theory of craft attends to the craft object's "practical physical function."[46] This function is not a mere abstract idea that precedes the object before it's made but a purpose the artisan grasps and works into the object, through technique, materials, and form.[47] Thus, Risatti explains, the "practical physical function is something inherent to and never imposed on an applied object. It is built into it by its maker and exists at the very core of the object as a physical entity, a formed piece of matter."[48] We see Hogarth, following Socrates, adopt a similar craft account of the way purpose is related to object.

In his final example of the chapter, Hogarth turns strategically to an example that does not seem to fit with the others: "The Hercules, by Glicon." What we now know as the *Farnese Hercules* hardly seems an object of utility. It does not fit the model of the shield, a beautiful thing that also has a specific function in the world, but nor does it fit the body like a chair. Hogarth provokes his reader with this example, suggesting that a classical statue, the very model for a conception of beauty as ideal and beyond substance, is related to the most pedestrian and functional objects. How is the Hercules like a chair or a racehorse? He introduces the statue by pointing us toward the first plate (fig. 1):

> The Hercules, by Glicon [fig. 3, plate 1], hath all its parts finely fitted for the purposes of the utmost strength, the texture of the human form will bear. The back, breast and shoulders have huge bones, and muscles adequate to the supposed active strength of its upper parts; but as less strength was required for its lower parts, the judicious sculptor, contrary to all modern rule of enlarging every part in proportion, lessen'd the size of the muscles gradually down towards the feet; and for the same reason made the neck large in circumference than any part of the head [fig. 4, plate 1]; otherwise the figure would have been burden'd with an unnecessary weight, which would have been a drawback from his strength, and in consequence of that, from its characteristic beauty.
>
> These seeming faults, which shew the superior anatomical knowledge as well as judgment of the ancients, are not to be found in the leaden imitations of it near Hyde-park. These saturnine genius's imagin'd they knew how to correct such apparent *disproportions*. (*AB* 26–27)

Hogarth's selection of the Hercules is no accident, for according to the modern, mathematical proportions he mocks in the preface (his example is Dürer, "fetter'd with his own impracticable rules of proportion") the

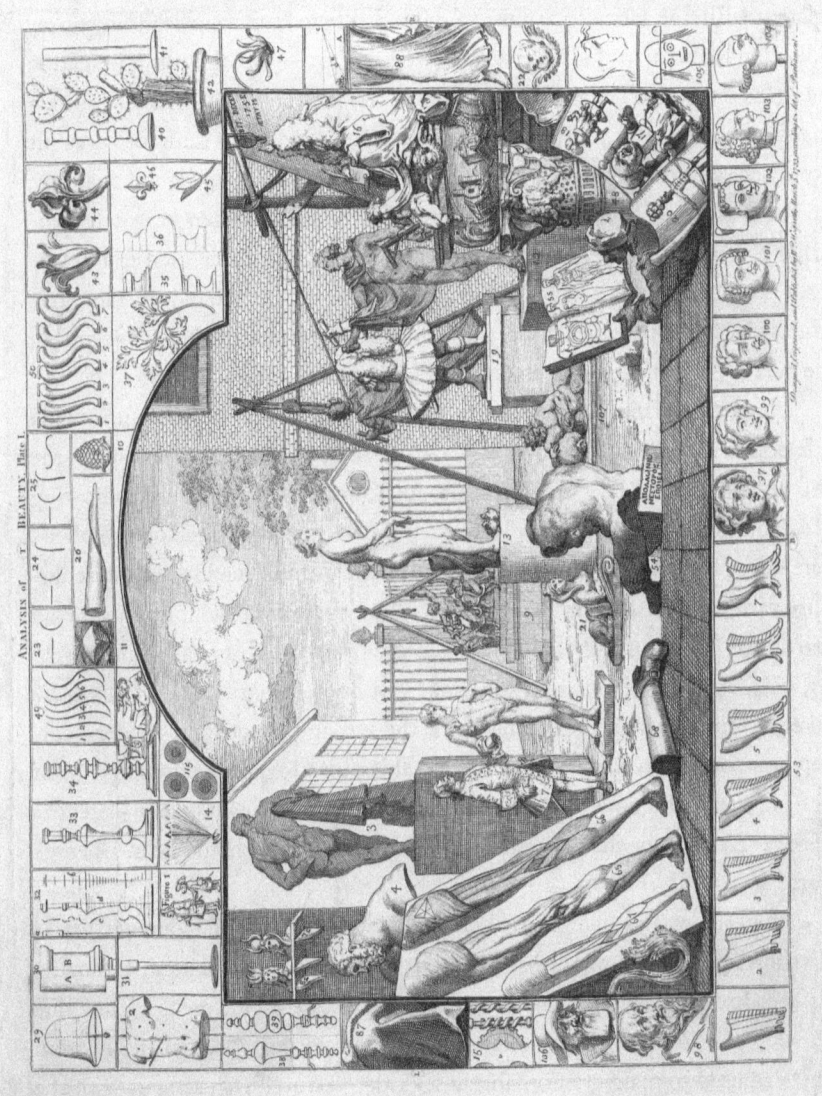

Figure 1. William Hogarth, *The Analysis of Beauty* (London, 1753), plate 1.

Hercules evidences "disproportions" (*AB* 5). Hogarth replaces mathematical proportions, then, with fitness: the statue "hath all its parts finely fitted for the purposes of the utmost strength." Now here Risatti and Hogarth part ways. Risatti would not see the Hercules as a craft object; rather, he would class it with fine art and attribute to it the ability to "communicate," to convey ideas.[49] Hogarth, for his part, wants to tie the two—use, the communication of ideas—together. Hogarth draws our attention to maker's knowledge as it pertains to the construction of the statue. This "judicious sculptor" uses his knowledge of the human form, indeed, his technical knowledge of anatomy, to see what is possible within the bounds of the body. When Hogarth identifies the purpose as "the utmost strength," he calls our attention to the way that proportions, once again, must adapt to the aim of the object. Hogarth, then, gives us "utility" or "fitness" as a means of considering how concept can be part of—built into—an object, even a beautiful statue.

Hogarth works this model of material form further in his use of the plate. He points us there at the outset of his discussion, offering us the statue on the left side of the tableau: "The Hercules, by Glicon [fig. 3, plate 1]." Following this notation, we look to the plate to see the back of the statue on the plate's left side. Now, we know that, although Hogarth took the idea of the sculpture yard from Xenophon, he represents a very particular sculpture yard on a particular corner in London: Henry Cheere's yard on Hyde Park Corner (*AB* 27, n41). And he tells us in the passage just quoted that in that yard we are likely to find "leaden imitations" with the wrong proportions, modern corrections for perceived ancient mistakes. This puts Hogarth's signal to consult the plate in ambiguous territory. For what are we looking at on the plate: the *Farnese Hercules*, or the distortion of that Hercules that should be in Cheere's yard?

Hogarth makes us aware of how form reaches into its context, just as he shows us how we might use utility to understand this relation. In having us see, in some sense, two statues at once, Hogarth forces us to imagine the *Farnese Hercules*, its proportions tampered with, its weight heavy from the lead used to make it affordable as a garden statue. Presenting us with the ideal form, its shape fit only to concept, Hogarth then follows up by having us imagine that form laden with contemporary associations such that we feel their presence in a changed shape and changed material. In terms of utility, this is not just about which example is "fit" and which not, but about how the material forms of the sculptures hold their purpose. Even if Hogarth hasn't smuggled it into "beauty," the context orientation of the relativism that worried Hutcheson is well established here, and it comes to us as an aspect of the artisan's material, formal production.

The Artisan's Plates

If Hogarth's simple account of sight seems like a weak tribute to Bacon, his embrace of the artisan does not. Indeed, Hogarth from the beginning of the treatise styles himself the kind of artisan-philosopher we recognize from Bacon's own account of natural philosophy (and Sprat's after it). The focus on fitness is a part of this, and Hogarth (like Socrates) encourages us to conceive of this attribute in artisanal terms, as something that can be understood as built into the object. Hogarth explores his handicraft philosopher potential in another register, too, as the example of the Hercules begins to suggest: as the maker of the engravings.

Hogarth begins the preface by styling himself a practitioner, in possession of a "practical knowledge of the whole art of painting (sculpture alone not being sufficient)" (*AB* 2). Later in the preface, he reinforces this idea in expressing his desire to write as an artist who has never taken up the "pen": "I . . . applied myself to several of my friends, whom I thought capable of taking up the pen for me, offering to furnish them with materials by word of mouth" (*AB* 13). We might neglect, or suppose incidental, this reference to ideas as "materials," the "matter or substance from which a thing is or may be made," were it not for Hogarth's subsequent lines.[50] He describes himself as having instead "thrown [the matter] into the form of a book" before submitting it to the judgment of his friends (*AB* 13). Here, "thrown" almost certainly means "to put into another form or shape" (a sense of the word that, in eighteenth-century discourse can refer to everything from translating language to tilling a field).[51] Given the declaration of Hogarth's practical knowledge, it is also hard not to hear "throw" in another contemporary sense: as "to form or fashion by means of a rotary or twisting motion"—descriptive of the manufacture of both pottery and silk.[52] Hogarth as a writer, he seems to remind us, is still Hogarth the artisan and future member of the Society of Arts, the written treatise at home with a pot or a piece of cloth.

In an earlier draft of the preface, Hogarth extends his thinking about the importance of acting out of "my own sphere."⁵³ Here, he describes his failure to pass the project to an "able pen" in even more thoroughly material, artisanal terms:

> and so like one who makes use of signs and jestures to convey his meaning, in a language he is but little master of, I, as an expedient, to make up for my deficiencys in writing, have had frequent recourse to my Pencil. Hopeing, that as the mechanick at his Loom is as likely to give as satisfactory an account of the materials, and composition, of the rich Brocade he weaves (tho uncouthly) as the smooth Tongue'd Mercer with all his parade of showy silks about him I may in like manner, make myself tolerably understood, by those who are at the pain of examining my Book, and prints together.⁵⁴

From the book as a manufactured, artisanal object, Hogarth takes us here through a complex set of metaphors to explain his justification for writing. He places himself in the position of "the mechanic at his Loom" and offers his knowledge of the "materials, and composition, of the rich Brocade" as on par with that of the mercer. In this description he makes plain that he sees the *Analysis* as though it is part of a Baconian "History of Trades" project, a how-to manual of a kind. But Hogarth has more to say about what kind of thing he is making. At first the modern reader might puzzle over Hogarth's account of compensating for his "deficiencys in writing" by "recourse to my Pencil." But of course he is drawing attention here to the difference between writing in ink and drawing, between the written part of the treatise and the plates, the "prints" that he claims are necessary for a "tolerable" understanding of the project. We might note Hogarth's insistence here on thinking about writing and drawing merely in terms of instrument and material, not in terms of words and images, stressing their connection, pulling us once again into an artisanal account of both. We take note, too, of his comparison of drawing (and ultimately engraving) as akin to "jestures" and their relation to a spoken language he does not know well. The artisan's body plays across the materiality of writing and drawing, and in prints that are like movements of the physical body. Engraving is a particularly interesting place to put these bodily movements, as engraving is often thought, historically, to have been about the fixing of the image.⁵⁵ Here, Hogarth turns that on its head, placing both movement and the body into the image itself.

Hogarth's "explanatory plates" are as central as this last quotation suggests, and he makes this clear in addressing them immediately in his published introduction. Now, it is worth remarking up front that the plates require some explanation. After all, privileging one's "own eyes" and drawing on Bacon to emphasize that we should search for the line in nature might leave the reader at a loss as to why Hogarth would include representations of beauty at all. The plot thickens in Hogarth's decision to include in plate 1 the depictions of multiple canonical classical statues routinely used in the instruction of the artist (and, indeed, visited by those despicable connoisseurs on the Grand Tour). Hogarth was already known for his decision at St. Martin's Lane Academy to have artists draw from live models, which both separated him from French Academy practice and generated controversy. With this in mind, then, we find Hogarth explaining that the "explanatory plates" do not contain the examples we might expect to find there:

> My figures, therefore, are to be consider'd in the same light, with those a mathematician makes with his pen, which may convey the idea of his demonstration, tho' not a line in them is either perfectly straight, or of that peculiar curvature he is treating of. Nay, so far was I from aiming at grace, that I purposely chose to be least accurate, where most beauty might be expected, that no stress might be laid on the figures to the prejudice of the work itself. (*AB* 17)

Hogarth clarifies that his figures are not "placed there by me as examples themselves, of beauty or grace." Ronald Paulson takes this to be another version of Hogarth's statement that we must "see with our own eyes."[56] In this reading, Hogarth tells us to examine the engravings only provisionally and to prepare to set them aside for real-life cases. This is the Hogarth opposed to French instruction by copying, the empiricist who wishes to guide his reader to examine nature only.

Two elements of Hogarth's description, however, send us back to the plate, rather than turning us away from it. First, he attends to the "figures" and "lines" of the mathematician, no neutral figure when it comes to a practical, experience-based treatise. We recall that Aristotle's theory/practice binary—and many versions that follow—employs mathematics to exemplify "theory." Indeed, the Euclidian geometrical lines with which the mathematician could be thought to work are "abstract, conceptual, rational," and have "neither body nor colour nor texture, nor any other tangible quality."[57] Not so the lines of Hogarth's mathematician, who works with a "pen" and whose lines are written on paper, full of body and texture, their inexactitude a mark of the human hand that "made" them.

Hogarth, barely out of the gate in explaining his plates, has turned us back to lines and to the maker, even toward usefulness. And what at first seems like an account of mere inexactitude, should resonate differently in this artisan's aesthetics. We should focus on Hogarth's positive account, on how the handmade lines "convey the idea of his demonstration," and we should puzzle over the sense in which we could take these lines to be analogous to Hogarth's own. When Hogarth pushes us toward contemplating the mathematician's "demonstration," he wishes to account for the way that lines might work in the way Reviel Netz describes for ancient "diagrams."[58] Netz, in his work on Greek mathematics, explains that diagrams are not like pictures. They are, rather, in his terms, *"psychological objects."*[59] They contain, as Hogarth suggests, a "subset of the real properties of the ... object."[60] In the impossible "perfectly straight" lines (Hogarth) and in the "so-called equilateral triangle" (Netz), diagrams have the status of "make-believe."[61] In this sense they are "functionally identical" to the intended object.[62] Diagrams work on what Netz calls an "ontological borderline"—gifted with a kind of reality that works within a "cognitive process" as the thing it represents.[63] Netz writes, "The diagram is not a representation of something else; it is the thing itself. It is not like a representation of a building, it is like a building, acted upon and constructed."[64]

Rather than separating out the representation from nature, then, Hogarth lands on a metaphor that encourages us to see both as ontologically similar. Moreover, he links this reality not just to the diagram but also to the hand. He abstracts the "idea of demonstration" and attaches it to diagram (as in Netz), while at the same time he doubles down on the handwriting of the diagram, a reality separate from its cognitive work. We can ponder the move again in relation to Hogarth's engraving. What, after all, could he be asking us to do in experiencing the plate as one experiences this rough geometry? Unlike a geometrical diagram, Hogarth's plate sets out to depict things in the world: a smoke jack, some stays, a human face, the *Farnese Hercules*. But here Hogarth, as in his comparison of the plate to embodied gesture, seems curious to have us reconsider its ontological status. What exactly is this thing?

Where are we supposed to look on it (in it?) to figure that out? And what does that mean for our way of perceiving the object in the first place?

There is more. For Hogarth, in that same passage, also teasingly tells the reader that he may have messed with the lines in his representation on the plate: "Nay, so far was I from aiming at grace, that I purposely chose to be least accurate, where most beauty might be expected, that no stress might be laid on the figures to the prejudice of the work itself." This is quite a claim, because part of Hogarth's goal in the engraving is to show us the "line of beauty" in the way only a visual image can; after all, we are otherwise reduced to language of "serpentine" or "S-curve"—hardly a clear idea of what he is after. This is where Paulson's account has the reader turn away from the representation, but surely this peculiar statement of Hogarth's may send us as much back to the plate as away from it. Can we see the evidence of this tinkering? Is the line of beauty in the *Apollo Belvedere,* or on that tulip, a little less exact that it might be? This kind of playfulness is typical of Hogarth in the *Analysis,* as I have already begun to suggest. And here is the upshot: Hogarth has us think of the plates as being handled, far beyond those original strokes of the "Pencil." They are made, they are adjusted, we are encouraged to scrutinize the plates for the trace of the hand of the artist and engraver. In a roundabout way, we are made to see the engraving as a collection of lines—made by the hand or by the burin. Notice how similar this is to Hogarth's account of geometry and diagram. We might think back, quickly, to Hutcheson's preference for the theorems and other abstractions that would require the inner sense of beauty. To this, Hogarth reminds us that the geometrician himself writes with rough marks; he is another author-artisan, working, on Hogarth's account, with the "pen." Abstractions, examples, and printed lines all are made present to us as the products of bodily making.

Hogarth turns again to making and lines in the mental experiment he proposes for improving the artist's way of seeing the world:

> In order to my being well understood, let every object under our consideration, be imagined to have its inward contents scoop'd out so nicely, as to have nothing of it left but a thin shell, exactly corresponding both in its inner and outer surface, to the shape of the object itself: and let us likewise suppose this thin shell to be made up of very fine threads, closely connected together, and equally perceptible, whether the eye is supposed

to observe them from without, or within; and we shall find the ideas of the two surfaces of this shell will naturally coincide. The very word, shell, makes us seem to see both surfaces alike. (*AB* 21)

This shell method, in which we turn solid objects into a shell and that shell into "very fine threads" is a technique that Hogarth imagines his reader practicing until it becomes habitual. If we perfect this sort of seeing over time, the viewer will "gradually arrive at the knack of recalling them into his mind when the objects themselves are not before him," and this will be "of infinite service to those who invent and draw from fancy, as well as enable those to be more correct who draw from the life" (*AB* 22). Hogarth's mental experiment turns lines into "threads," transforming an act of abstraction into a process of making. If we pull this back to Hogarth's participation in the debates over the continental academy, we can see that he makes the abstract process of "design," the painterly idea of concept over object, into itself a material thing.[65] Moreover, the closer we look at Hogarth's language, the more it seems as though he is describing his own material, engraved lines. After all, lines made on copperplate with a burin must be "closely connected together, and equally perceptible," the space between them necessary to prevent the ink from pooling across lines and disturbing the image. There is much more to say about this mental habit that Hogarth hopes we might all acquire, but first I must turn to threads and lines themselves.

The Line

Now the line of beauty, Hogarth's most famous contribution from the *Analysis*, is often treated as a bit of an embarrassment. The singular and universal "line of beauty" has been tough to square with the deep cultural analysis of Hogarth's paintings and engravings. Ronald Paulson and Jenny Uglow, two of the artist's twentieth-century biographers, make no bones about the conflict they observe between Hogarth's art and his conception of beauty. Hogarth's "real mistake," writes Uglow, "was to defy the tyranny of rules by inventing a new rule himself, and insisting that it was an absolute truth."[66] Paulson, in his introduction to the *Analysis*, makes a related point: "Hogarth had introduced 'the Power of habit and custom' in his Preface to beat connoisseurs and their dupes. . . . He seemed unaware that the argument could be turned against his own Line of Beauty, which might be as locally and perhaps as ethnically conditioned."⁶⁷ The "mistake" can seem especially grating because it appears to toss aside what Hogarth accomplished in his more famous graphic prints, in their insistence that "beauty could be local and cultural," and indeed, in their insistence on ordinary circumstances (the poverty of Gin Lane or the life of a young prostitute) as the basis for art.⁶⁸

We can only understand how Hogarth brings these elements—line, everyday life—together by considering much further how he extends the body-based thinking about representation (including "utility") into his understanding of the line of beauty and of lines more generally. Let's begin with Hogarth's textual introduction of the reader to the line, one of the instances that must have frustrated the biographers. Here, Hogarth argues that he is not the creator of the line of beauty, merely its expert reader. Hogarth's minihistory of the line of beauty resembles in important ways his friend Henry Fielding's history of the "comic epic poem in prose" in the preface to *Joseph Andrews*, where Fielding argues that his new form of writing is both novel and identifiable in classical literature. Likewise, Hogarth offers us a history of the line of beauty as the history of an object hiding in plain sight. He begins the preface with writers on painting, isolating passages in Lomazzo, Charles Alphonse du Fresnoy, and Roger de Piles Piles, where "grace" is posited but not understood, resulting in the *"Je ne sçai quoi"* that Hogarth finds worthless (*AB* 4). Hogarth can do better, he claims, returning to a "precept which Michael Angelo deliver'd so long ago in an oracle-like manner," the identity of the line of beauty (*AB* 4). Hogarth's discovery was, then, anticipated by the great painter but has not been understood until now. Hogarth replicates this move for his own audience by placing the line of beauty in a self-portrait (and labeling it) before publishing his treatise. This, ultimately, can be understood as a rather sophisticated advertising ploy, but it also contributes to understanding the line as a symbol shrouded in secrecy. In the *Analysis* he describes this instance of the line as akin to an "Egyptian hierogliphic," gesturing toward the controversy and "amuse[ment]" it caused within his community of artists (*AB* 6). The line of beauty, then, begins as an enigma, and as a symbol that might also be a form of writing. It waits, in true Enlightenment spirit, to be rationalized in a treatise.

In such depictions Hogarth offers the line of beauty as a symbol to be decoded. But from its first appearance in the self-portrait, *Gulielmus Hogarth*, the artist playfully messes with what sort of thing the line is (fig. 2). Although

Figure 2. William Hogarth, *Gulielmus Hogarth* (London, 1749). Courtesy of The Lewis Walpole Library, Yale University.

he later refers to this as a "hieroglyphic," the line of beauty in the self-portrait is given to the viewer labeled and placed on the artist's palette. The line is rendered in three dimensions, its shadow visible on the surface of the palette, its ends lifting away from the surface. The viewer may already wonder, then, about the line: is it a symbol, even a form of writing, or an object? Hogarth stresses the line's materiality by placing it on the palette, next to his daubs of paint. But the line also has a similar sheen and angle to the burin that lies between the palette and Hogarth's pug. Could the line, then, be less a material to make with, more a tool with which to do the making?

Hogarth continues to play with the line's object-identity in two subsequent images: the subscription ticket, *Columbus Breaking the Egg*, and the image on the title page of the *Analysis*. In *Columbus Breaking the Egg*, the engraved ticket shows Columbus in the position of Jesus at the last supper, engaged in a common-sense refutation of the challenge to stand an egg on end (he crushes it to make the end flat).[69] An allegory of Hogarth's own common-sense answer to the problem of "beauty," the engraving depicts some cooked eels on a plate in the foreground. The two eels, displayed as lines of beauty, sit on a plate with an untouched egg. And Hogarth does not let go of the literalization of what is elsewhere called the serpentine line. On the title page of the *Analysis*, Hogarth gives us a serpent in a clear glass pyramid, the three-dimensional encasement accentuating the serpent's curve in space (fig. 3). The serpent has its own symbolic role to play, and we don't have to guess at it: Hogarth quotes Milton's *Paradise Lost* lines on Eve and the serpent on the title page. But he also tempts us to delight in how this symbolic role is only one way the line makes its meaning. After all, we have already seen the line as a smooth object in the self-portrait and as a cooked eel on a plate; on the title page it is transformed again, becoming a vivified—if also dead and preserved—serpent. The line crosses high and low (playing both the common food of the eel, as well as the high, symbolic Miltonic serpent). Throughout these visual representations, the line flaunts its malleable ontology: symbol and thing, abstraction and tool, it then performs a once-living animal and another one dead but revivified in its presentation. To make matters more confusing yet, Hogarth offers us the serpent in a kind of glass case, calling our attention to its resemblance to a natural history specimen.[70] But the species of this animal appears to be philosophical, aesthetic: the label on its case says "Variety." Living, dead, or lifelike, the encased serpent is

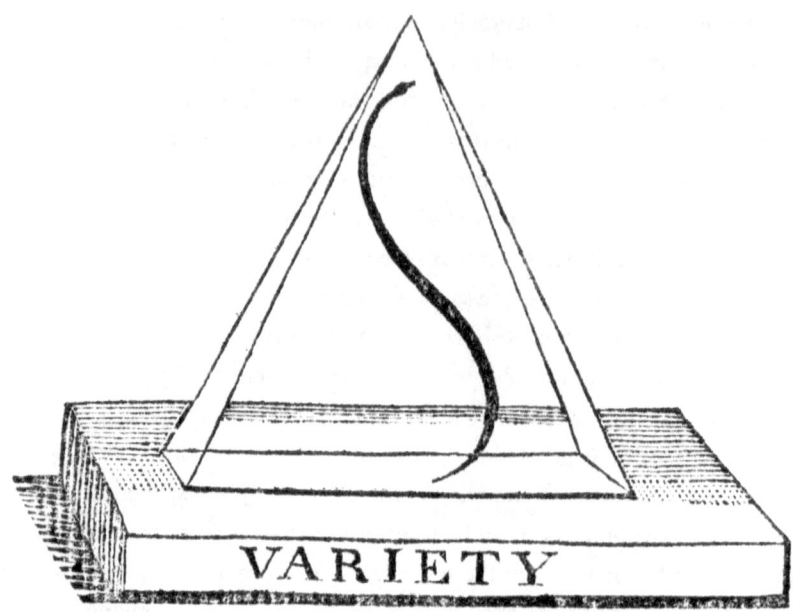

Figure 3. William Hogarth, detail from the title page of *The Analysis of Beauty* (London, Printed by J. Reeves, 1753). Courtesy of The Lewis Walpole Library, Yale University.

turned back into the aesthetic register, captured once again as line and as an aesthetic example.

Before we even reach the first chapter of the *Analysis of Beauty*, then, Hogarth insists we be aware that lines are more complex than we may have thought them to be. In this, he resembles the theorist of the line Tim Ingold, who begins one of his books on lines by reveling a bit in the strangeness of its subject matter. He quotes his academic colleagues' confusion and even slight dismissal. What kind of a thing is it: "Is it a theory? Is it a metaphor?" To these questions, Ingold will answer no: "Lines, I insisted, are phenomena in themselves. They are really there, in us and around us."[71] Like Hogarth, Ingold begins by hinting that the line might stand for something else. But that is only on the way to our entertaining it as fully real, only an introduction to our appreciating the dimensions of reality it could contain. In a provocative first sentence of his introduction, Ingold asks, "What do walking, weaving, observing, singing, storytelling, drawing, and writing have in common?" He answers that they all "proceed along lines," something we might have come to from Hogarth's own treatise.[72]

But how do they so proceed? Hogarth answers in part through considering "reality" in terms of a very specific form of pictorial reality that would have been on his mind as a maker of images. That is, Hogarth is interested in the way that pictures, two-dimensional works, posit a third dimension and how that speaks to the kind of reality they contain. We can see this in tracing a resemblance between Hogarth's examples and those of a later art theorist, E. H. Gombrich. For when Hogarth puts the snake heads on the serpent and the eels, he plays around with the kind of visual joke that fascinates Gombrich in *Art and Illusion:* one that demonstrates how a minimal addition can change radically the kind of thing we take the line to be. Gombrich, for his part, recalls a childhood fascination with "a simple drawing game I found in my primer":

> A little rhyme explained how you could first draw a circle to represent a loaf of bread (for loaves were round in my native Vienna); a curve added on top would turn the loaf into a shopping bag; two little squiggles on its handle would make it shrink into a purse; and now by adding a tail, here was a cat. What intrigued me, as I learned the trick, was the power of metamorphosis: the tail destroyed the purse and created the cat; you cannot see one without obliterating the other.[73]

Like Gombrich, Hogarth has a theorist's interest in the play of line and in the simplicity of its radical transformations (as well as in children's art and line art).[74] But unlike the later theorist of art, Hogarth does not at all seem confident of Gombrich's punchline for the circle-to-cat exercise: "you cannot see one without obliterating the other." That is, Hogarth seems to go out of his way to make sure that we aim to see both at once by labeling the serpent as a line. As we plunge into issues surrounding what Gombrich calls the "psychology of perception," Hogarth's remark that we should "see with our own eyes" now seems almost tongue-and-cheek in its simple presentation of this world wild with illusions and visual tricks.

Gombrich helps to contextualize Hogarth's shell experiment: it is both an exercise for the artist's vision and another reality experiment, along the lines of the serpent. Like Gombrich, Hogarth is sure painting involves training the painter's hand and the "painter's eye."[75] Gombrich writes of those who teach the art student to "find means of battling down his knowledge of the familiar meaning of things and look only at shapes and tones projected onto an imaginary plane."[76] He quotes Cézanne's advice to Bernard, "to look at nature in terms of simple shapes of known property, that is, in terms of cylinder cones, and spheres," which aims to "increase [his] awareness of pure shapes and relationships."[77] Hogarth's own shell method works like this and he means it to be "of infinite service to those who invent and draw from fancy, as well as enable those to be more correct who draw from life" (*AB* 22).

Hogarth sees fit to train all his readers, even those of us who will continue without painting. Jonathan Kramnick expands on that generality to draw our attention to Hogarth's scrutiny of our acts of perception in the manner of a psychologist of perception or phenomenologist. Using artisanal metaphors, Kramnick draws out Hogarth's perceptual trick through a relation to the philosophical tradition's connection of thinking to handwork. In Kramnick's reading, Hogarth's experiment shows that "the artwork's end is to put the beholder in place . . . to, as it were, handle everything she sees."[78] This phenomenological account reads Hogarth's account as revelatory about the way we might perceive reality generally, and actively: "both to pursue and to compose what comes into view," to "draw its lines and to move with" the world around us.[79] Kramnick draws his model for active perception from a source also privileged in Gombrich's psychology of perception: J. J. Gibson, the psychologist of perception.

Hogarth's shell, however, takes us another way and pushes back at the ideas of reality grounding Gibson, Gombrich, and Kramnick. For in Hogarth's model the artisan's hand doesn't just offer us a description of how perception works in the world around us; rather, it draws attention to a reality problem that concerns pictorial representation.[80] For inside Hogarth's shell there is another experiment along the lines of the serpent head and the circle-to-cat game. The passage is intriguing in part because it makes us repeatedly transform our thinking about the object. Thus, although Uglow, for instance, reads this as a kind of graphic modeling (avant la lettre), Hogarth forces us to see the shell in discrete stages.[81] It is first solid, then hollow ("scoop'd out"), then a "thin shell" (*AB* 21). But this is not enough for Hogarth, who insists that the shell "be made up of very fine threads, closely connected together, and equally perceptible." Hogarth unpacks the object, and he forces us to think and rethink the material abstraction that we have in mind. As part of this, he plays with how we think of the surface, moving us from a surface that is solid (it's "thin") to something not does not seem a surface at all, made of threads that he says we can still see individually, "closely connected together, and equally perceptible" (*AB* 21).

Here, Hogarth seems to draw on a passage from the very beginning of Leon Battista Alberti's *Of Painting*, which begins with lines and points as geometrical entities. Alberti writes, "These Points extended on in a Row close to each other form a line: And therefore the Line, with us, shall be a Mark whose Length may be divided into Parts; but at the same Time it shall be so thin, that it can never be split. . . . Many Lines conjoined together after the Manner of Threads in a Piece of Cloth, form a Superficie."[82] Hogarth, obviously, is not so sure about this account of surface or about its fluid slide from geometry to cloth. Hogarth works on the metaphor, arranging the strings in order to draw attention the strangeness of the resultant "Superficie" (surface) of threads. Ingold, who also finds himself intrigued by the same passage, provides a reason why. As he puts it, "A thread is a filament of some kind, which may be entangled with other threads or suspended between points in three-dimensional space. At a relatively microscopic level threads have surfaces."[83] Any craftsman would be aware of such surfaces, of course. And Ingold reminds us—as does Hogarth—that threads are not just materials for making (and not just material lines) but themselves made things. To Alberti's geometry, Hogarth's elaborate making and remaking of the shell answers that there is no simple surface created by placing these "very slender" things with their own surfaces, side by side.

Hogarth wishes us to see threads and shell together, and Gombrich helps us to see what is at stake in doing so. Gombrich puts his childhood circle-to-cat drawing game before the reader early in *Art and Illusion*, because it provides an account of the way our minds perceive the reality of pictures, paving the way for Gombrich's fuller account of pictorial illusion, how a picture of two dimensions creates for us three. Working with J. J. Gibson's perceptual psychology, Gombrich grounds his theory of illusion on what Robert Briscoe terms the "*Continuity Hypothesis*"—that is, Gombrich argues that "pictorial experience is continuous with the experience of seeing face-to-face."[84] Gibson himself says of a picture that it is "both a scene and a surface, and the scene is paradoxically behind the surface."[85] This perceived "behindness" encourages us to recognize what's in the picture as another version of our own reality.

For this model of visual perception to work it is important that perception itself does not tolerate ambiguity. As Briscoe explains it, "The visual system selects only one consistent, 3D-scene-interpretation of the image at a time, even when more than one such interpretation can be made to 'fit.'"[86] Readers of Gombrich will recall that he is insistent on this feature of perception in the opening pages of *Art and Illusion,* and he repeatedly calls up visual examples that allow us to prove it for ourselves. The circle-to-cat example is one, along with his characteristically emphatic "you cannot see the one without obliterating the other." He proves the same thing to us with the more famous duck-rabbit example, and he insists on it further with an example of Kenneth Clark's: the inability to see Velásquez's "brush strokes and daubs of pigment" and his "transfigured reality" at the same time.[87]

In a reading of Hogarth through phenomenology, the hand works to extend reality, whether in the world in front of us or in the work of art. In Hogarth's double vision of reality, however, the hand provides its own distinct reality, forcing us to see the picture as a made thing. As I have already suggested with the serpent's head, Hogarth seems interested in Gombrich's puzzles, even as he repeatedly suggests a doubleness that Gombrich's model will not hold. In the case of the shell, Hogarth tells us how to break the object down, and the written passage makes this breakdown linear. This is not a rabbit-duck image with which we might struggle. Rather, he gives us a solid object, requires us to dig it out as we would a material one, calls it a shell, asks us to reduce it and reformulate it as threads, and then calls it a shell again. Even when Hogarth writes, "The very word, shell, makes us seem to see both surfaces alike," one senses that he is grappling with a duck-rabbit situation, turning to the word ("The very word shell") to insist that all is still well in this layered object of one surface and many. My point is not that Hogarth gets it wrong or, indeed, that Gombrich does. Rather, Hogarth does his best to show how something might cut into the very device that should create "continuity" and create a reality of its own. That he does this while ostensibly telling us how to see is typical of the creative riddles Hogarth poses for us in the text of the *Analysis.*

In the shell, Hogarth offers us a theorization of the art object as handicraft, one that should call up too his insistence on his hand in the printed engraving. As a visual artist, Hogarth offers us a theory of how these two realities might coexist, and he dramatizes it in prose and in the plates of the treatise. When Briscoe sorts philosophical explanations of the duck-rabbit example, he gives us an alternative to Gombrich that is closer to Hogarth's aims, a position that he associates with the philosopher Richard Wollheim: "When we look at a picture, we enjoy an experience as of the depicted, 3D scene. This experience, however, is always fused with awareness of the superficial pattern on the pictorial surface. . . . In this respect, pictorial experience has two different dimensions or "folds" of representational content."[88] For Hogarth, the constant attention to the artisan's hand and its trace on the "surface" on the work of art accomplishes this doubleness of representation that is also a doubleness within perception. While Gombrich would say that one could see only the daubs of paint or the reality of the image, Hogarth has us see both. Hogarth might have been especially attentive to this double possibility in his imitation of the Dutch painters that Svetlana Alpers describes as concerned with the surface of their own paintings as finely crafted things.[89]

We can see another performance of this relation if we pull away from Hogarth's prose and look at the first plate. Here again, Hogarth considers how the hand of the artisan could tip the reality of the painting by urging us to see it as an art object. Both "explanatory" plates present a tableau surrounded by smaller pictures in what appears to be a kind of internal frame. Each tableau—sculpture yard and country dance—presents a scene that seems to work with "continuity theory"; that is, it seems to be a place that corresponds to our own reality with the scene behind the surface, as Gibson describes it. Frames usually protect that reality, marking off the inside of the painting from the outside of the world around it. Indeed, the frame can make the illusion easier to see, its spatial analogy crisp. The frame may also be decorative, the work of the craftsman but likely not of the artist. Consider, then, Hogarth's work with the internal frame. For one thing, he pulls the frame into the world of the engraving, permitting the border to be part of the work itself. He also fills that border, not with the ornate decoration that might be found in other kinds of internal frames, but with examples of the line of beauty.[90] Some of these examples are fairly straightforward for the viewer, showing us how to pick out the beautiful line in a series of seven examples, for instance, or showing us natural forms in which the line appears. But some of these inset pictures are quite a bit more complex. One, in the middle right of the frame, offers a tiny drawing on perspective that Hogarth tells us in the treatise he hopes may compensate for those readers "who have not learnt perspective" (*AB* 28). Another shows us the process for remaking a candlestick design, leading us through stages of rejiggering the stick's proportions, the arrangements of its lines and curves. These "examples" are not just examples of the line; they are snippets of how-to manuals, and they draw the reader into the form of rationalized how-to knowledge that Celina Fox associates with the "useful arts" and Gilbert Ryle with "knowing how." Hogarth, as he will remind us in the tableau itself, had just manufactured a frontispiece to such a volume on perspective, his friend Joshua Kirby's book *Dr. Brook Taylor's Method of Perspective Made Easy, Both in Theory and Practice*.[91]

How-to manuals, those bastions of "theory and practice," demonstrate maker's hands-on and process-based knowledge. In staging the plates with such knowledge in the frame, Hogarth argues that the supplemental might define the artwork itself, even as he offers a sort of metaexample of the daubs of paint on the artist's canvas. The immediate question these little diagrams pose is the one with which he has been experimenting: what kind of a thing is this engraving to begin with? As a further part of this visual argument, Hogarth offers an account of perspective in the tableau—and not too far from the tiny drawing on perspective. Here, Ronald Paulson has observed, Hogarth offers a visual trick that's familiar from his frontispiece on false perspective for Kirby's book, where he offers some perspectival jokes, showing us the kind of humor possible if one does not maintain a consistent viewing position for the reader of the print.[92] By flattening perspective Hogarth inserts a bit of action into the first plate (in similar fashion), where the statue of Apollo seems to hit a statue of Brutus on the head, while Brutus stabs Julius Caesar (the statue hanging suspended from a rope) in the back. Right in the middle of his illusion, Hogarth calls attention to his own hand *and* to his practical work on perspective. When he makes a tiny pocket of the illusion read as flat, he places that in conversation with the other how-to knowledge on the plate. Once again, he highlights—here on a larger scale—how the two folds of reality might be present and the kind of practical hands-on knowledge that might fill one of them.

Hogarth's complex visual riddles continue to invoke the hand as the key to the way representations contain reality. Over and over, one reality gets tucked inside another, and lines help us to see how that is so. In my earlier example of the Hercules and its surroundings, in the shell's threads, and in the tableau's frame, Hogarth shows us the way apparent continuity can contain yet one level of reality more. This was part of his extraordinary presentation of the line in the first place as a piece of writing, an animal, a specimen. How might it be more than one of these, he asks us, at once?

One reading of Hogarth's examination of the line over the course of the treatise says that he considers this universal sign or symbol as it moves through history. There it is in Glykon's statue, in Michelangelo, in a sculpture yard down the street. But by now we must surely acknowledge that this is not a mere symbol moving through history. It has more to say about reality, too, than the word I first entertained for the sculpture yard: "context." After all, what Hogarth encourages us to imagine is that making is part of the thing and a reality attached to it, even if some version of that thing discounts that entirely. Put another way, the particularity Hogarth shows us, starting with the Hercules, does not dissolve into the universal line; it coexists with it. And it does so, because he has shown us the line's ability to contain more than one reality.

If this seems incoherent from our more modern perspective, we should place Hogarth's experiment alongside the account of "taste" (Hogarth's own word on the title page) that James Noggle gives in his examination of eighteenth-century taste in terms of temporality. Noggle argues that taste has within it two apparently incompatible temporal accounts: one of "intense immediacy" and universality; the other of "the long process," taste as a product of "evolved, collective processes and outcomes."[93] Noggle explains that the eighteenth century tolerated what seems to us like a radical incompatibility: these "two different facts about taste" were "to be reconciled (or just accepted) in different ways, according to different exigencies."[94] If we take Noggle's model to Hogarth, then, we can see that Hogarth's suspension of universal and particular might be ordinary for this period, even if his formal means of demonstration is not.

Action

In concluding, I look at Hogarth's final chapter on "Action." This chapter allows us to see Hogarth's continued experimentation with form and society, as he contemplates in different terms how society might make its way into representation. In some ways this argument takes on the same kind of challenge as "utility," building concept into form. In this final case, Hogarth makes this formal point by telling us what we as readers of the treatise already know: that lines are made by hands, always embodied, and that their reality is changeable, expandable.

In this last chapter Hogarth begins by separating the beautiful from the everyday by thinking of them in terms of two kinds of lines. There are two kinds of movements, he says, "useful" and "graceful," and only the latter expresses the line of beauty. All "useful habitual motions, such as are readiest to serve the necessary purposes of life, are those made up of plain lines" (*AB* 106). As for graceful movements (and their waving lines): "The whole business of life may be carried on without them" (*AB* 106). But this is a distinction that does not hold, as well it should not for a theory that elsewhere ties beauty to utility. In his account of the graceful movement of the arm in the "habit of moving in the lines of grace and beauty," Hogarth seems invested (much like Locke is in his treatise on education) in thinking through how such movements become customary: they "by frequent repetitions will become so familiar to the parts so exercised, that on proper occasion they make them as it were of their own accord" (*AB* 106, 107). The customary seemed to separate straight and beautiful, but now all is customary. This repetition (not unlike the working on silver or making objects into threads) gets the habitual, the customary, into action and, indeed, into the line of beauty itself. What Hogarth has been proposing from Hercules onward now seems given to us as process, as though in slow motion, as bodily habit and custom is made into a line.

As in the example of polite gestures, Hogarth is quite clear that the manners of everyday eighteenth-century British life get pulled into these lines too. Hogarth begins with simple actions: "bodies in motion always describe some line or other in the air, as the whirling round of a fire-brand apparently makes a circle, and the water-fall part of a curve, the arrow and bullet, by the swiftness of their motions, nearly a straight line" (*AB* 105). But this is not the only way action can make its way into form, as Hogarth makes clear in turning to "*habit* and custom" (*AB* 105). The lines that "describe" an individual's "gait in walking" for example, stand not just for those actions (of lifting the foot or swinging the arms) but for "the habits [each person has] contracted" (*AB* 105). Lest we imagine this just as an imaginary line, a line marked in the air, Hogarth fills out the picture by turning us to material lines on the page:

for as it is with walking, so it is too, with the "visibly different" handwriting that marks everyone's habitual movements (*AB* 105). That motion that the line contained when made "with pen or pencil" was never just abstract shape; it was always attached to the body and to habit. As we saw in Hogarth's contemplation of himself as a mechanic weaver, writing, drawing, and weaving provide examples of maker's knowledge. And they all, as Ingold reminds us, move along the line.

We have seen Hogarth push making and acting into the line; here we see him offer instead how "*habits* and customs" have been making lines all along. Hogarth doesn't go too far with this, but he forces us to pursue it for a moment, when he treats dance. The minuet, he writes, is really an intensification of the "ordinary undulating motion of the body, in common walking"; moreover, "[t]he figure of the minuet-path on the floor is always composed of serpentine lines"—something that can be said, as well, of the very different country dance (*AB* 109). Movement, habitual movement, can be tracked for the individual or for the group. Habit, then, is broadened to this shared experience: walking becomes dancing, the shape of which "var[ies] a little with the fashion" (*AB* 109). Hogarth's second plate bears out this interest, showing us the country dance as a dynamic set of social relations. Hogarth goes further yet in his characterization of these dancers by comparison with other "barbarian" dancers: "[t]he dances of barbarians are always represented without these [graceful] movements, being only composed of wild skiping, jumping, and turning round, or running backward and forward, with convulsive shrugs and distorted gestures" (*AB* 111).

Hogarth ends up in a crude binary, but this should not distract us from the larger implications of his argument. Having set out to define a universal line of beauty, Hogarth has ended up somewhere else, where lines signal something about the societies of which they are a part. They can do this because of the kinds of representations they are: able always to be themselves and something else, and to come by their representational status by way of a connection to the body. Lines can pull together "content" and "abstraction," but they do so by destroying the relation between the two.

Long after anthropology's formulation as a discipline, philosophers have continued to use the line and, indeed, the line as an embodied mark, to consider how to accomplish the aim of integrating theory and practice. Philosophers whose aim is the proper understanding of society, that is, repeatedly turn to representation to work out theory and practice in formal terms. We can see this in *The Practice of Everyday Life*, where Michel de Certeau begins by describing "the signifying practices" of consumers, who act with "artisan-like inventiveness."[95] De Certeau encourages us to imagine this inventiveness in terms of Fernand Deligny's "wandering lines" depicting the everyday movements in space of autistic children, what de Certeau calls "'indirect' or 'errant' trajectories obeying their own logic."[96] The strange "tracings" that children and caretakers alike map into everyday spaces serve for de Certeau as a way of thinking about how "everyday practices" can be "tactical in character," potentially directed against structures of power.[97]

By the time de Certeau finds his lines, he has perused not only Deligny's book on autism but also earlier ideas about representation within anthropology and sociology. Bourdieu had already called attention both to "practice" and to representation as a problem: the trouble, in his words, of "constitut[ing] practical activity as an object of observation and analysis" is also the trouble of the analyst introducing "into the object the principles of his relation to the object."[98] Bourdieu's terms are familiar. In the first pages of *Outline of a Theory of Practice*, with structuralism in mind, he observes that the "social world" is the object of "three modes of theoretical knowledge which have only one thing in common: the fact that they are opposed to practical knowledge."[99] In this project to define further what exactly such practical knowledge would entail, Bourdieu casts the relation between the theoretical and the practical in terms of lines: the "logical" relationships are those shown by lines on a "map," and the practical relationships are shown as "a network of beaten tracks made ever more practicable by constant use."[100] Like Hogarth, both Bourdieu and de Certeau fight the limitations of representation not by turning away from it entirely but by attempting to find new ground within representation itself. Bourdieu does not want to jettison structuralism altogether, but he does want to demonstrate that structuralism's map of relations contains additional relations that are undeclared.

When de Certeau and Bourdieu turn to formal resolutions for a theoretical problem, they turn to practice: to working out the problem on the page. De Certeau's aim for doing this can leave us with a new set of questions about Hogarth. De Certeau's example reminds us, for instance, of the origin of Hogarth's line: both in the rococo style, originating in France but popularized by practitioners at St. Martin's Lane (like Hubert-Francois Gravelot, the engraver, who was famous for his book designs), and in graffiti and children's art.[101] The rococo line was seen as "errant"—empty, counter to reason—at least partly due to its "perceived origin in the craft workshop"; indeed, even its proponents in Britain were middle-class practitioners, not architects and their patrons.[102] The rococo line, then, stood for design that had emerged through craft, and the vitality of Hogarth's lines could reach toward this wandering, antisystematic potential. Hogarth forms his way of thinking about society on this edge, and this brings together the subjects of his paintings and engravings and his aesthetics. The form he privileges, and the way he teaches us about it, move against the system, gauging "politeness" from the margins.

FOUR

Science

PRACTICING SCIENCE IN BANKS'S TAHITI

JOSEPH BANKS JOINED CAPTAIN COOK's *Endeavour* voyage (1768–71) as an amateur botanist, but his journal from the voyage is best known for its "ethnographic" reporting.[1] As J. C. Beaglehole, the modern editor of the journal, describes Banks's trajectory over the course of the voyage, "the natural historian becomes the natural historian of man."[2] I argue in what follows that we have not appreciated the complexity in Banks's turn toward reporting on the societies of Tahiti and New Zealand: what it captures about the use of science in the "contact zone" and what it can tell us about the history of anthropology. In the context of this book's argument, Banks performs in Oceania a turn we have seen before in fiction, when Mr. Spectator takes his natural philosophical viewing to London society. But this is not London, and Banks examines the potential for craft knowledge to bridge the difference between the society he has left and the one into which he and the *Endeavour* crew intrude.

Although readers have jumped to read Johnson's and Boswell's journals as "philosophical," the same has not been true for Banks's. There is good reason for this. The *Endeavour* voyage was the first British professional scientific voyage of its kind, and Banks travels on this voyage as a scientist and member of the Royal Society. Although his journal is not the place where he records plant specimens—this happens in a separate set of descriptions—the journal often follows both maritime and natural philosophical conventions. Like Cook's journal, Banks's was to be turned into the Admiralty at the conclusion of the voyage, and Banks clearly models his own entries both on

Cook's journal and log and on the many printed travel texts that accompanied both men on the voyage. Like seamen for over four hundred years, as well as Royal Society scientists for the past hundred, Banks writes in the "plain stile" and clearly assumes that his journal will have an audience of both casual, armchair readers and future voyagers.[3]

As we have seen in Defoe, however, the "plain stile" can do more than its prescribed referential work. Indeed, its referential capacity can even pave the way for an openness to experience that makes its way in around the edges. On this scientific voyage Banks attends to the methods prescribed by the natural philosophy of the Royal Society and by the natural history of Linnaean botany, even as he consistently thinks beyond the narrow accounts of experience specified by these scientific models. In Banks's hands, the plain style bends to accommodate a range of experience that both factual reporting and Linnaean taxonomy set aside, and craft knowledge is used to grasp this range. In the final section of the chapter, I look at Banks's representations of objects in Tahiti and the intensification of his distrust of writing to represent them. Moving toward small hand drawings, Banks attempts to connect with the world around him. As he does so, though across the globe in Tahiti, he has a copy of Hogarth's *Analysis* at his disposal.[4] Whether or not the botanist read the treatise, when Banks reaches for society, he reaches for the hand and the line, showing us that a knowledge of society may begin by interrogating the process of representation.

The *Endeavour* voyage was the first professional British scientific voyage. Coming after the close of the Seven Years' War with France, it also was an unambiguous attempt to fill out Britain's knowledge of the South Atlantic and Pacific, setting the groundwork for further imperial expansion and competition with other European powers. In this chapter I examine Banks's use of craft knowledge as a part of this imperial project. Banks's efforts sometimes quite overtly disrupt the forms of scientific knowledge that have been considered most violent, most attached to the "imperial eye." In this sense, I follow a train of recent scholarship in finding Banks's a less authoritative, less certain, imperialist text than earlier frameworks would lead us to expect.[5] I am interested too in the way Banks goes beyond the "vulnerability" or "weakness" in the imperial gaze; the forms of experience he employs here at times look "ethical" in more modern terms. Banks's wide variety of ways of engaging with his environment, I argue, should inspire some caution in how we, as critics and historians, mobilize later philosophy of experience for our own ethical ends.

Crafting Perception

In early April 1769, Captain James Cook and his crew aboard the *Endeavour* are hunting the seas for their destination, Tahiti. In the days before the ship pulls into harbor, those on the ship see a series of islands. "Land in sight," Joseph Banks, the naturalist on board, writes repeatedly, as they pass by the islands Marokau and Ravahere, and Reitoru.[6] The observer's obligations in terms of these sightings are clear: particular description and coordinates that would allow "future navigators," as Banks puts it, to identify the land masses on future voyages. Especially without a reliable way to calculate longitude, the descriptions of land and coastline were vital, and we see both Cook and Banks use earlier printed descriptions to orient themselves. For his part, Cook gives us the terse description we would expect of the captain of a naval vessel. On the approach to Marokau and Ravahere, his entry from Friday, April 7, begins:

> *Winds East. Courses N 66° W. Distce sail'd in miles 66. Latd in South* 17°48'. *Longd in West of Greenwich* 143°31'. Fresh gales and Clowdy. At ½ past 2 pm got up with the East end of the land seen yesterday at Noon, and which prov'd to be an Assemblage of Islands join'd together by Reefs, and extending themselves NWBN & SEBS in length 8 or 9 Leagues and of various breadths, but there appeared to be a total separation in the Mid[dl]e by a Channell of half a Mile broad, and on this account they are called the *Two Groups*.[7]

Banks's account of these islands is listed on April 6; he gives his account in equally plain language that, as we might expect, is markedly less technical but committed to description that would be of use to future travelers:

> Pleasant breeze, at ½ past 11 land in sight again, at 3 came up with it, proved to be two distinct Islands with many small ones near them Joining by reefs under water.
>
> The Islands themselves were long thin strips of land ranging in all directions sometime ten or more miles in lengh but never more than a quarter of a mile broad; upon them were many Cocoa nut and other trees and many inhabitants . . . (B 1:246)

As is typical of this section of the journal, Banks describes the situation of the islands, estimates their size, and, in what follows describes the behavior, weapons, and physical appearance of the islanders—all detailed markers of identification for later voyagers.

By the time of the *Endeavour* voyage, the plain style is a virtually unshakeable convention for the traveler, especially the scientific one. Almost a hundred years earlier William Dampier opens his bestselling *New Voyage Round the World* (1697) by calling attention to his "plain piece" as having the "stile" belonging to a "Seaman," even as he dedicates the volume to Charles Montague, the president of the Royal Society.[8] Dampier's double move calls attention to the plain language that had been the rule for mariners for centuries, even as he shows his awareness of that style having been made new again with the Royal Society's endorsement. As for the dedication to Montague, the feeling was mutual. The Royal Society praised Dampier in the *Philosophical Transactions*, and, generally, seems to have viewed him as a success story in the Baconian campaign to make "seamen . . . the agents of science."[9] In the following century, things would be mainly as Dampier has them, with travel writing, along with science, committed to facticity grounded not in the scholar's style of writing but, as Sprat tells us, in that of the "merchant or tradesman," or mariner.

Dampier is a closer match to Banks than the intervening century might suggest, because the conventions of the scientific traveler were exceptionally durable. As Paul Smethurst puts it, for most of the eighteenth century, traveler-scientists such as Cook and Banks were still guided by empirical protocols laid down in the previous century."[10] The Royal Society enlisted Robert Boyle to write "General Heads for a *Natural History of a Countrey, Great or Small*" in 1666, which followed Bacon closely in advising travelers to order their findings under certain categories or "heads."[11] This standardization allowed for Bacon's dream of a many-handed scientific effort that might amass a large body of knowledge; it also, as Jason Pearl argues, aimed to "control the act of representation," telling travelers what to look for and how to understand it. In his account Pearl encourages us to see these reports as akin to scientific instruments that "disembody the senses."[12]

Cook's entry from April 7 reminds us that he is responsible for more than just plain description. Rather, Cook's two major tasks for the journey, as specified by the Royal Society and the Admiralty (cosponsors of the trip) are related to mapping. The trip's most obvious scientific mission is the recording of the transit of Venus, which would—along with several other such observations—provide data that could allow the calculation of the earth's distance from the sun.[13] The trip's other mission, this one a kind of open secret, was the investigation of a possible Southern Continent or *Terra Australis*, which had been hypothesized for decades and whose importance had been newly revived by the Royal Society Fellow, Alexander Dalrymple. And, of course, Cook was to chart and map whatever he could. He came to the voyage as a former marine surveyor; he had charted the coast of Newfoundland, a qualification that the Admiralty held in high esteem. Indeed, Cook chose to do the surveying work on the voyage himself, even though a captain usually would have handed off such work to a lower surveyor. Cook's skill, as Beaglehole tells us, made surveying "almost natural . . . skill in translating experience on to the drawn and written page."[14] Copies of Cook's charts were delivered to the Admiralty along with his journals; the original charts show that sixteen were marked out for publication, and Hawkesworth's controversial publication of the journals. Contained most of these charts, though there are over thirty unpublished charts and maps.[15]

Above all else, then, the *Endeavour* voyage trafficked in what Bruno Latour calls "immutable mobiles," the "signs, prints, and diagrams" that defined new "ways in which groups of people argue with one another" by preserving their shape as they circulated.[16] Print culture's rendering of scientific facticity in terms of two-dimensional visualization allows what Latour calls "hard facts" to become harder yet, strengthened through comparison and addition; these images create a new way to do natural philosophy and, indeed, a new way to perceive the world.[17] Banks, the naturalist, has his own immutable mobiles, bringing back not only plant specimens but drawings of those specimens: the plan, from the outset, was not just identification but circulation.[18] "Artificial curiosities," objects from the cultures Banks visited, were likewise both collected and, ultimately, engraved. (The output is significant enough that Latour cites the presence of documents from Cook's voyage on the voyage that is his own central case, La Perouse's Pacific voyage.)

I'd like to have Latour's "immutable mobiles" in mind in looking at another form of description that Banks provides. On April 9, three days after the description of the pair of islands, Banks writes of an unusual day during which the ship has not passed a new island. He begins:

Fine weather and pleasant breeze. It is now almost night and time for me to wind up the clue of my this days lucubrations, so as we have found no Island, I shall employ the time and paper which I had allotted to describe one in a work which I am sure will be more usefull at, if not more entertaining to all future navigators, by describing the method which we took to cure Cabbage in England; which Cabbage we have eat every day since we left Cape Horne and have now good store of, remaining as good at least to our palates and full as green and pleasing to the eye as if it was bought fresh every morning at Covent Garden market. Our Steward has given me the receipt which I shall copy exactly false spelling exceptd. (B 1:248–49)

Banks's full transcription of the recipe follows:

Take a strong Iron bound cask for no weak or wooden bound one should ever be trusted in a long voyage, take out the head and when the whole is well cleand cover the bottom with salt. Then take the Cabbage and stripping off the outside leaves take the rest leaf by leaf till you come to the heart which cut into four; these leaves and heart lay upon the Salt about 2 or 3 inches thick and sprinkle Salt pretty thick over them, and lay cabbage upon the salt stratum super thick till the cask is full. Then lay on the head of the cask with a weight which in 5 or 6 days will have pressd the cabbage into a much smaller compass. After this fill up the cask with more cabbage as before directed and Head it up. N. B. the Cabbage should be gatherd in dry weather some time after sun rise that the dew may not be upon it. Halves of cabbages are better for keeping than single leaves. (B 1:249)

We have gone some distance by the end of this passage, from island observation to the "heads" and "single leaves" of salted cabbage. Such wild leaps are of course part and parcel of travel writing at this historical moment, organized by date and "teeming with disparate kinds of information."[19] Yet this passage is exceptional, because Banks calls our attention to the jump, to the substitution. He tells us that what he is about to relate will occupy the same "time and paper" that he had reserved for an island's description. He makes light of this, to be sure, even toying with Horace's "delight and instruct" to let us know that this will be more "usefull" and "entertaining," both.

The tone is playful, but, as the allusion to Horace indicates, Banks means for us to consider what the text is trying to do for its reader. Describing his journal in terms of "space" and "paper" reminds us of the critical medium of distribution of Latour's immutable mobiles, and, in fact, conjures up the space an island might take on a map. Its more useful substitute is another, possibly compatible, form of natural philosophical knowledge: the recipe. The recipe, associated with the artisan and with early natural philosophical experimentation, certainly continues the knowledge project of the voyage. Everything is fair game for exploration: from a new land mass to the process of concocting a new vegetable prevention for scurvy. And the recipe, too, can be understood as an immutable mobile, in that it abstracts from practice a repeatable list of substances and steps. Like maps, recipes can travel across the globe, as Banks dramatizes here.

Natural philosophy does not separate the forms of knowing that Banks urges us to distinguish with his proposal of substitution. In the "*Hints* offered to the consideration of Captain Cook, Mr Bankes, Doctor Solander, and the other Gentlemen who go upon the Expedition on Board the *Endeavour*," James Douglas, fourteenth Earl of Morton and president of the Royal Society, makes no such distinction (C 1:514–19). He proceeds seamlessly from asking the men to find out the existence of a Southern Continent, to asking them to report on "*Mechanics*, Tools, and manner of using them," should they find people living there (C 1:516). Baconian reporting could leave off here. But, for his part, Banks seems to pry open the possibility that we recognize from other accounts of craft knowledge in this period: its connection with the body. Banks relies on the reader to identify this difference. For while the map (or a description of the physical presence of an island) calls on us to see that a particular island is there, located in a particular point on the globe, the recipe calls us do something—or at least to follow along with the actions for doing and making. Recipes are at the root of Baconian experimental philosophy, William Eamon tells us, and they show the presence of the artisan in that later natural philosophical world.[20] Eamon helps us to see how the recipe can encode process even in its stable, listlike form. As he puts it, "Recipes are the record of trial-and-error experimentation. They are the accumulated experience of practitioners boiled down to a rule. We trust recipes because we know that behind them stands someone who does not use them."[21] Banks is not this original craftsman, operating only through the body, but he does establish his own practical relationship to the text of the recipe. "Our Steward has given me the receipt," Banks explains, "which I shall copy exactly, false spelling exceptd" (B 1:249). If the recipe, or "receipt," is a kind of abstraction of practice, Banks insists on a very material response to that abstraction: copying it by hand—and with improvement, subtle transformation—into the text of the journal. Banks draws his own journal writing into the contrast he has asked us to establish between "knowing that" (the island, the map) and "knowing how" (the recipe).[22]

Banks shores up this distinction between kinds of knowledge when he refers to wrapping up his journal entry as "wind[ing] up the clue of my this days lucubrations" (B 1:248). Here, Banks likens his presentation of the recipe (which ends that day's journaling) to winding up the "clue" or the end of a ball of yarn. Again, the emphasis is on the journal's distance from the visual and its turn toward a tactile maker's knowledge. He has been writing and also, Banks suggests to us, he has been weaving a story or following a thread. It could be that the recipe is just an outgrowth, almost a doubling, of the kind of work that has been doing all along. Much as Locke transforms his metaphor of mind from the *Essay*'s blank page to *Education*'s crafted or "wrought" object, Banks writes only to call attention to that writing as a form of making. The recipe may take up space, and indeed may occupy that space with writing; but its action keeps returning us to the body and to process.

As he substitutes recipe for island, Banks is on his way to the place where the *Endeavour*'s group of scientists will observe the transit of Venus. Tahiti is the *Endeavour*'s destination because it is the one place in the "South Seas" that the Royal Society knows Cook can locate with relative ease. The *Dolphin*'s return in the spring of 1768, just months before the *Endeavour*'s voyage, meant that this particular island would not need to be "re-discovered" (as some islands first visited by Europeans in the sixteenth century would need to be) in order to serve as a scientific base.[23] Sailing toward Tahiti, then, meant sailing through rather uncharted seas on the way to the one point on the map that had already been established as known. And, sure enough, when Cook reaches Tahiti and notes its coordinates, the first thing he does is to compare those coordinates with Wallace's. They are a close match.

But as the ship nears Tahiti, nears this point of cartographical near-certainty, Banks takes another route in his own descriptions. When, a day after his entry on salted cabbage, on April 10, the ship finally approaches

Tahiti (or, in the journal, "George's Land" or "Georges Island"), Banks remains very much interested in the process through which this important destination comes to be known:

> This morning the wind from N to NW, the weather very hazey and thick. About 9 it cleard up a little and showd us Osnabrug Island discovered by the Dolphin in her last voyage, it was distant about 6 leagues and appeard like a very short cone. Very light winds NW. About one land was seen ahead in the direction of Georges Land, it was however so faint that very few could see it. Soon after it was seen off the deck in the same faint manner but appearing high. Our distance when it was first seen was 25 leagues. At sun set the ship was nearly abreast Osnabrug Island 2 or 3 leagues from it, it apprd to have many trees upon it but in some parts the rocks were quite bare.
>
> At this time it remaind in dispute whether what had been so long seen to the Westward was realy land or only vapours; myself went to the Masthead but the sunset was cloudy and we could see nothing of it. (B 1:249–50)

What the men see through the hazy sky is indeed Tahiti. For his part, Cook does not report these false starts; rather, he offers the following day, "At 6am saw *King Georges Island* extending from WBS½S to WBN½N it appear'd very high and Mountainous" (C 1:73). As the view becomes clearer, Banks notes, similarly, "Georges Island in sight appearing very high in the same direction" (B 1:250). But even then, his interest seems captivated less by the arrival of the island than on how it had been perceived: "I found the fault was in our eyes yesterday tho the non-seers were much more numerous in the ship than the seers" (B 1:250). As land nears, as the potential for description increases ("Land in sight!"), Banks, rather, focuses on "our eyes." He moves back to the old perceptions of the island (or its potential), and he consigns the men to be named for their old perceptions of it: "seers" and "non-seers." They embody their perceptions; the turn is not outward toward the world but toward the "eyes" of the bodies perceiving it.

Banks turns to the body in another way, too, in the following paragraph, when he returns to cabbage.[24] As they sail nearer to their destination of seven months' pursuit, Banks again digresses and again points us to the "paper" on which that digression is recorded: "I shall fill a little paper in describing the means which I have taken to prevent the scurvy in particular" (B 1:250).

His concern is still cabbage and methods of preparing cabbage (at least at the start), but the how-to form moves into something closer to a formal scientific experiment. While Cook tidies up his arrival by noting latitude and longitude and by comparing his position to the earlier measurements by the crew of the *Dolphin*, Banks gives us recipe and experimental method. He relates that he has tried both fermented and salted cabbage (as per the previous recipe), but that he then, experiencing only partial success, moved on to the "lemon Juice which had been put up for me according to Dr. Hulmes method describd in his book and in his letter which is inserted here" (B 1:251). This time, we have both the body that makes the lemon juice and the body that receives it. We learn of the amounts of each substance Banks has ingested; we hear of his swollen gums and "some small pimples . . . in the inside of my mouth which threatned to become ulcers" (B 1:251). And we hear of their improvement, down to the "few pimples" left on Banks's face (B 1:251). If his first description of cabbage gestured toward the body in the form of the recipe, this passage goes further. Here, the body serves as the source of the cure, even as its blemishes bring its experience to the fore. About to depart the ship, Banks puts our minds less on Tahiti, more on gums and pimples.

At the moment when sight should triumph—"Land in sight!" or "There it is!"—Banks hits the reader with the contingencies involved in seeing at all, and moves us toward a focus on the body. Certainly we are quite distant, here, from the vision-based realism that has been associated with the travel writing genre in this period, as are we from the "disinterested scientific inquiry" or the "objective" stance which has been assumed to be that of the scientific traveler. We are closer, to be sure, to the "practical reason" that Margaret Cohen describes in *The Novel and the Sea* and that she associates with Cook and the mariner in particular.[25] But we can see how Banks is forcing us to new questions about the "plain stile" that Cohen describes as one of the mariner's "tools."[26] It could be that Cook comes by this kind of reporting rather automatically, by way of his profession, as Cohen and Beaglehole suggest.[27] But Banks does not. He, rather, is engaged in attempting to find something within natural philosophy that registers such practicality, and he does so by setting one form of scientific representation over and against another (all in the plain style): geographic location versus how-to.

Making a Specimen

Banks reflects on the breadth of experience assumed by different forms of scientific inquiry, asking us to compare the cartographer to the artisan or experimental philosopher. In the pages of the journal focused on the earliest days in Tahiti, Banks turns this same kind of interrogation on his own practice of botanical taxonomy. By the time Banks came to propose occupying a place on the *Endeavour*—something orchestrated through his friend the Earl of Sandwich and only subsequently approved by Cook—Banks had learned and taught himself a great deal as an amateur botanist. He remarked in a letter to Thomas Pennant of 1768, "Although I am not a pupil of Linnaeus, however I know his method, & reckon myself to be a kind of Linnaean being."[28] On the voyage Banks stocked his retinue with other men attached to the great naturalist: Dr. Daniel Solander, a Swede and favorite student of Linnaeus's (he had contributed plants to the Linnaean herbarium) who planned to send specimens back to the Master, as well as Herman Sporing, whose father in Finland was a professor and correspondent of Linnaeus's, and who was himself most likely an able naturalist, though employed by Banks as a kind of secretary. Like Linnaeus himself, the men would not draw their own specimens. This task fell to Sydney Parkinson, one of Banks's draftsmen. Parkinson was trained in the commercial world of botanical cloth design, and he would be devoted almost exclusively to natural history specimens.

Since Foucault's account in *The Order of Things*, Linnean taxonomy has come to stand for the Enlightenment's "exhaustive ordering of the world."[29] Taxonomy is characteristic of what Foucault calls the Classical Age, because it sees the "individual" in terms of a "table of visible differences": "It defines, then, the general law of beings, and at the same time the conditions under which it is possible to know them."[30] Mary Louise Pratt has shown us how such a taxonomy functions on the ground and in the "contact zone." As she puts it, "Natural history called upon human intervention (intellectual, mainly) to compose an order," an order that, from the very outset, was global and the product of the "(lettered, male, European) eye."[31] Pushing out of the way "local knowledges," taxonomy renders the unfamiliar familiar, or even renders it the taxonomist's own, revealing nature to be a kind of collection.[32] Natural history is extractive and straightforwardly dehumanizing, "subsum[ing] culture and history into nature" and neglecting the "functional, experiential relations among people, plants, and animals."³³

Aboard the *Endeavour*, botany is a major enterprise. Banks and Solander collect specimens (doubtless along with Banks's servants), which are dried, painted, and "described" in Linnaean terms. After Banks returns to England, the botanical collection ultimately would to consolidate his reputation as a member of the Royal Society. As a knowledge system, Linnaean botany is crisp, clean, and totalizing. For Foucault it brings together the "knowledge of nature" and "a philosophy of representation."³⁴ Classification, he tells us, "requires only words applied, without intermediary, to things themselves."³⁵ Foucault implicitly cautions us that Sprat's goal for words and things does not merely operate in referential terms, where a word is plain enough that it might be mistaken for a thing. Rather, such an aim operates, too, in the tabular presentation of natural history, where words—names—precede things, lay the groundwork for them. In natural history, "the thing itself . . . appears in its own characters, but within the reality that has been patterned from the very outset by the name."³⁶

We see this kind of predictive naming, this naming that is always already organizational, in Banks's journal. Out of Plymouth for less than twenty-four hours, Banks makes his first Linnaean identification. His entry of August 26, 1768, relates a "fair" wind and "light breezes"; he says he "saw this Even a shoal of those fish which are particularly calld *Porpoises* by the seamen, probably the *Delphinus Phocaena* of Linnaeus, as their noses are very blunt" (B 1:153). Here we find Banks using Linnaean taxonomy to correct the common seaman's term. But later in the voyage, his descriptions of identification are shortened, as sight and taxonomy are collapsed: "Shooting again, killd *Nectris munda* and *Procellaria saltatrix*" (B 1:467). This shorthand creates the effect of Banks seeing taxonomically, identifying the bird as—or even before—he kills it, a conflation Lorraine Daston and Peter Galison refer to as "synthetic perception."³⁷ There is no identity for the bird—in the natural world, in another lexicon—beyond the taxonomy. This is a world in which, as Foucault puts it, there is "language already imprinted on things."³⁸

But the journal itself is not a scientific atlas like the ones Daston and Galison consider in their prehistory of objectivity. Quite the contrary, it is devoted to the much messier consolidation of such images—and here, too,

Banks is interested in representing the reach of experience. Soon after the *Endeavour* arrives in Tahiti, Banks describes Parkinson's labors in painting a specimen:

> The flies have been so troublesome ever since we have been ashore that we can scarce get any business done for them; they eat the painters colours off the paper as fast as they can be laid on, and if a fish is to be drawn there is more trouble in keeping them off it than in the drawing itself.
>
> Many expedients have been thought of, none succeed better than a mosquito net which covers table chair painter and drawings, but even that is not sufficient, a fly trap was nesscessary to set within this to atract the vermin from eating the colours. For that purpose yesterday tarr and molasses was mixt together but did not succeed. The plate smeard with it was left on the outside of the tent to clean: one of the Indians observing this took an opportunity when he thought that no one observd him to take some of this mixture up into his hand, I saw and was curious to know for what use it was intended, the gentleman had a large sore upon his backside to which this clammy liniament was applyd but with what success I never took the pains to enquire. (B 1:260–61)

In this scene of painting, the taxonomic process of abstraction, taking a fish or a plant and placing it in a table, is turned open to the environment. The result is not just destructive—of specimen and painting alike—but explosive for the senses. Foucault writes of the way that natural history "leaves sight with an almost exclusive privilege."[39] Even within sight, the eye must pick and choose, structuring observation to produce the right kind of object. In this operation, the senses are pared down; what is left for sight entails "a visibility freed from all other sensory burdens"—even color becomes "lines, surfaces, forms, reliefs."[40] Banks's description of the taxonomic scene, by contrast, shows visibility under all kinds of sensory burdens. In a nightmare for Foucault's taxonomer, "Colour" is no mere distracting hue in this description but an extraordinarily material pigment, doubly excessive to the form of the natural object. For "colour" not only adds hue to image; it is a physical substance that provokes an especially unseemly sensory response, one associated with "lack of certainty" and "variability": taste.[41] If taxonomy, as Pratt warns us, can disturb the local systems on the ground, here specimen and representation alike—as well as the act of knowing they constitute—are consumed by their environment. Banks stresses this in placing flies and painter in direct competition: flies "eat the painters colours off the paper as fast as they can be laid on" (B 1:260). One order competes with another (the one supposed to regulate it), the natural world consuming the usually unacknowledged materials of the taxonomic one.

But in this world of dissolving taxonomy, of representation reclaimed by nature, another kind of scientific pursuit seems right at home, a kind of knowing that privileges nature in all its dynamism: experimental philosophy. The natural history scene of Banks's relation becomes the occasion for a cascade of experiments. From net to fly trap, we follow the experimental attempts to contain the flies—or to attract them. As any good experimentalist would do, Banks documents failures along with successes. Indeed, the second comparison to the painter does not turn to the flies but to the flycatcher: "if a fish is to be drawn there is more trouble in keeping them off it than in the drawing itself." As Banks is about to reveal, "keeping them off" is strategic and practical; it is also hand work, stringing nets and making pastes, "trouble" or work that is set up in his formulation directly against the "drawing itself." Far from the streamlined forms and abstraction that natural history covets, Banks renders both interactions with the natural environment, painting and flycatching, as craft.

The scene of painting a natural history specimen rejects the clarity of the taxonomic table and lands in process. In Banks's description, though, this seems to produce not closure but incredible productivity: in experiments and in what Latour would call actants. The scene ends with a Tahitian man who has taken the flycatcher ointment to use as a medicine for his backside. Let's review what happens in this part of the scene: the flycatching "mixture" of tar and molasses is unsuccessful in luring the flies and is set outside the tent for disposal; at this point, the "Indian" comes along, seizes an "opportunity," and places the unknown substance on his "sore." The descent of the scene into process reveals that there are a lot of different kinds of actors behind the painting of a specimen. Daston and Gallison point out that in Linnean taxonomy there is almost always "four-eyed sight," a combined effort of the taxonomist or natural historian and the painter: one to see and formulate, the other to see and draw.[42] Banks and Parkinson share this "four-eyed sight," but Banks reveals the scene to be made by many more actors yet, all of them as invisible in the final product as he is. The figures that dominate this passage are the flies who "consume" the paint, the actors (it is unclear which) who attempt to catch the flies, and the Tahitian man who takes the mixture. Banks lets us see that getting fish or plant to paper includes all of these actors. It is true, of course, that Banks seems to disparage and mock this "Indian" at the end of the passage, calling him a "gentleman" and telling us that he, Banks, didn't follow up on the cure, implying that it is too insignificant for him to bother. But the reader cannot but be struck by the fact that Banks has given us the vocabulary to describe and comprehend this particular action as experiment. Banks show us his own strategy with the cabbage and the lemon juice, and his formulation leaves us with something of a puzzle: What, after all, is foreground and what background here? Is the Tahitian man's experiment part of the taxonomic exercise, or could it be the other way around, with taxonomy the prelude to and productive of the materials for an experiment on the body?

Banks does not help us to resolve this matter, but he does show us an example of what happens when representation opens to craft: not just a turn from object to process but a disruption of the singularity of "maker." This, he suggests, might be true for craft and for that thing called "experiment."

Craft and Performance

Banks's consideration of experience as grounded in craft and determined through experiment is expansive. I have examined thus far how this appears in his descriptions of knowing nature—through a map, through a taxonomy—and his attention to the way embodied, craft knowledge exceeds these ways of knowing. I want to turn now to the way that craft knowledge figures in Banks's attempts to know Tahitian society. The episode I have in mind is a famous one in which Banks participates in a mourning ceremony. In reference to this scene Banks is sometimes called a "participant-observer," a term indebted to a later moment in anthropology.[43] But the anachronistic "participant-observer" is an oxymoron that contrasts ideas in which we have already seen Banks take interest: it depicts a position at once defined by its remove, its distance from the thing or action, and by its involvement, its central role, in that action.[44] This later solution to the problem of theory versus practice within anthropology is in tune with some of the limitations we have seen Banks contemplate in theory-driven cartography and natural history. Even if we cannot call Banks a participant-observer, exactly, we can follow this lead in attending to the very deliberate way in which he considers participation—and his curiosity about what kind of knowledge will result from it.

Let's look at the ceremony into which Banks insinuates himself. Here, the historical knowledge is thin and European, since Tahiti was at that time an oral culture. Banks's is the first written account of the ceremony that the anthropologist and historian Douglas L. Oliver calls "heva-tuppa'u," the "mourners' masque" or "ghost's night."[45] Oliver attempts to consolidate accounts spanning from Banks through an observation in the 1920s. The six later European accounts Oliver considers reiterate the basic division Banks marks out for the participants in the ceremony, between a "principal mourner" and "attendants."[46] The ritual, Oliver observes, began daily and lasted for as long as three weeks or a month; over time it would diminish in frequency and then be abandoned. In it, as he explains, a "band of masked mourners probably made their rounds every day, starting and ending with obsequies at the ghost-house, and scoured the countryside for about an hour or so."[47] The ceremony would have been striking, in part due to the elaborate dress of the principal mourner (Cook returned with one, and it remains in the British Museum) and in part for its violence and action, for the chief and his attendants all carried weapons—clubs and spears—with which to threaten or injure members of the community who were not quick enough to escape them.

Banks begins his reporting as an observer, struck by an initial performance of the ceremony. He tells us that the heva-tupapa'u or "most singular dress" draws his attention and leads him to approach Tubourai, the lead mourner, to ask if he may accompany him. Turbourai says that Banks must "consent[] to perform a character," and Banks agrees to do so on the spot (B 1:288). The following day, as he explains his entrance into the group of mourners, he notes that "Tubourai was the *Heiva*," and "the three others and myself were to *Nineveh*." Banks's Tahitian vocabulary here is inexact,[48] but Oliver's analysis of later reports makes clear that Banks is separating out the two kinds of participants in the ceremony. Banks then reports the movements of the mourners, from the "Corps" to Tubourai's own home (the occasion for Tubourai "praying twice"), then toward the fort, toward the shore, and "into the woods" (B 1:288–89). In all places, the "Indians . . . fly before the *Heiva* like sheep before a woolf." Banks makes clear his part in this frightful scene: "We the *Ninevehs*" are ordered to "disperse" the people (B 1:289).

For good reason the scene has been understood in terms of the theatrical metaphor that Banks provides, when he points out that he will "perform a character." As Vanessa Smith has shown, this metaphor is consonant with the way in which "sympathy" was understood in the period, as well as with our more modern suspicion that Banks was playing a part he had no cultural right to play.[49] Yet despite Banks's powerful metaphor, the terms in which he ultimately describes his participation cut against the actor-role division that we associate with European theater. Let's take a look at the way Banks introduces us to the ceremony. He begins by telling us that Tubourai "put on his dress" (B 1:288). At this point, gesturing toward the reader, Banks turns us toward a drawing that must have been interleaved in the journal's pages: "the figure annexd will explain it far better than words can" (B 1:288–89). When he comes to his own preparation, however, Banks opts for language and describes a process quite distinct from putting on a "dress." Rather, as Banks relates it,

> I was next prepard by stripping off my European cloths and putting me on a small strip of cloth round my waist, the only garment I was allowd to have, but I had no pretensions to be ashamd of my nakedness for neither of the women were a bit more covrd than myself. They then began to smut me and themselves with charcoal and water, the Indian boy was compleatly black, the women and myself as low as our shoulders. (B 1:289)

If we expect a costume of some sort for Banks, he suggests that we consider the transformation otherwise. Rather than dress, he recounts "stripping" and "nakedness." The replacement for his "European cloths" is "a small strip of cloth" that, Banks says, those helping him were "putting me on." The awkwardness of this last phrase, which seems a partial reversal of "putting on me," points to the uncertainty, here, of the priority of body or cloth. Likewise, as Smith has observed, "strip" and "stripping" seem to cancel each other out (as do "cloths" and "cloth"), raising again the possibility of clothing that somehow results in nakedness.[50] This stress on the naked body points to Banks's transformation as different in kind from Tubourai's, and different, too, from an actor's performance of a character. Whereas Tubourai dons a dress (and mask), transforming him into the chief mourner, Banks insists on a process that first pares down cultural markers and then makes body and clothing indistinguishable from each other. Gone is an easy distinction between the body of the actor and dress of the character that is critical for the way we ordinarily consider theatrical "character."⁵¹ Banks, rather, insists on an identity that lies on the surface of the body and with an uncertain distinction from it.

What does it mean for Banks to "participate" or "perform" in the ceremony? It's worth pausing to reflect on the awkwardness of this question for the passage at hand. For participation, here, seems to have reconfigured the self on which our models of participation and performance would seem to rely: the passage doesn't give us a psychological "self" that could draw back from the body (an intention separate from it), and it renders the body as a made thing. This objectification continues as the passage proceeds. Banks has already told us that he was "prepard" by others, and this preparation turns out to involve more than just the "stripping" mode of dress; it also entails a recognizable act of making. Banks relates, "They then began to smut me and themselves with charcoal and water, the Indian boy was compleatly black, the women and myself as low as our shoulders" (B 1:289). Here Banks reports the process of bodily coloring as a kind of craft, specifying its ingredients as "charcoal and water," thus recalling his own cabbage and lemon juice remedies and the molasses and tar flycatcher. But "smutting" is a bit different from the application of these other substances to bodies, for to "smut" often does not require a human actor: it may refer to the residual particles smoke, for instance, leaves on a wall.⁵² Banks describes the process of bodily marking by calling up a process that foregrounds substance over human actor. And he renders his own body the crafted object, rather than the body of the maker. To "perform," then, involves rendering your body an object that is worked on by others. And this holds through the rest of the ceremony, where Banks's actions in the ceremony are decidedly not his own; he merely follows the orders of Tubourai, the chief mourner, who tells the attendants where to move and what to do.

Banks describes not only a transformation of his body in this episode; he also describes a transformation in what a body—and indeed a person—can be thought to do. Take the consequences of the bodily "smutting." Banks observes quickly that his "smutted" body draws him into relations with women and children: "the Indian boy was completely black, the women and myself as low as our shoulders." He even assesses his own nakedness by way of "the women," neither of whom "were a bit more coverd than myself" (B 1:289).

Marks on bodies, even naked bodies, seem to confound ordinary social distinctions rather than accentuate them. And bodies are sorted, reconfigured in this way, based on what happens in or on the skin.[53]

If the first connections visible in this passage are between the "how-to" of ceremonial preparation and other recipes that Banks recounts in the language of experiment, then the next set of connections visible are between this passage and Banks's descriptions of an indigenous craft: tattooing. In his account of tattooing in Tahiti in particular, Banks is fascinated by the way that what he calls "marks" are placed inside the body, "under their skins" (B 1:335). The closest Banks can come to a metaphor is "inlaying," a practice of embedding one substance inside another, usually so that it is flush with the surface (B 1:335).[54] Banks gives a summary account of how this is done—what technology and materials are used—in a kind of recipe form. He also contemplates the social significance of these markings, speculating that Tahitian practice might be related to individual "humour"—or to age (due to particular ceremonies he has observed), or to "superstition" (B 1: 336). Observing the practice in New Zealand, Banks wonders if tattoos might reflect "different circumstances of [the person's] life" (B 1:335). In short, Banks contemplates a crafted social identity that takes things his own European society would have considered internal, like belief, or would be hard to pin down, like "circumstances," and rendered them visible, on the near-surface of the body.

The much later anthropologist Alfred Gell, working in the same part of the world Banks visited, calls tattoos a "double skin," and he draws attention to the inside/outside conflation that should recall Banks's "stripping": "the ink is absorbed into the interior of the body . . . but [is] still visible."[55] Describing something very close to the logic Banks works out in the passage on the mourning ceremony, Gell puts it this way: "The basic schema of tattooing is thus definable as the exteriorization of the interior which is simultaneously the interiorization of the exterior."[56] This back-and-forth bears out the logic of the exteriorization of the self through the objectified body that Banks has already shown us.

The account of character-as-tattoo shows that Banks has begun to consider two craft practices side by side (or even as one): European scientific experiment and Tahitian tattoo, which are entangled in the passage's explanatory logic. Postcolonial historians have rightly considered European approaches to craft in non-European societies as extractive. And Banks embraces such extraction on terms that Bacon would have applauded, as when he describes dye that might be useful for British cloth manufacture or recipes for preparing breadfruit, which might inform future colonists on a central part of their diet. Banks also brings back to England material plant specimens and "artificial curiosities" from the voyage, some of which remain in modern museum collections. Even as he literally extracts objects from Oceania Banks also uses craft—including Tahitian craft—to consider the parameters of experience. What shows up in his account of tattooing, I'm suggesting, goes well beyond natural philosophy's ordinary purview, because it is interested, not only in how something is made, but in how making can show us something about the relationship between one person and another, even one person and the society or group of which he is a part.

At many points, Banks and Cook are pointedly obscure about the value of indigenous knowledge. Historians have remarked for decades, for instance, on how peculiar—and disturbing—it is that Cook, for instance, has almost nothing to say about Tahitian navigational processes, even though he lived among these seafaring people for three months. Historians of practice, however, have brought our attention to the way in which collaboration between European and Tahitian navigators not only existed aboard the *Endeavour* but is contained in documents from the voyage, chief among them the map known as "Tupaia's map," after the master navigator from Raiatea. Nicholas Thomas describes this map as an example of "merged" knowledge systems.[57] The European, two-dimensional, Mercator-projection map also contains Tupaia's Oceanic navigational knowledge, grounded in memory and "structured on orientation and directionality," rather than placement on a grid.[58] Like Western navigators, Oceanic navigators use the stars, but that astronomical knowledge is part of an "integrated bod[y] of natural knowledge," a "dynamic cognitive map" that is "almost entirely mental with very little manifestation."[59] The components of this cognition, as it is understood through ethnographers, involve a version of a star compass; *etak*, a mental technique for estimating the distance to be traveled; and a strategic approach to island finding that "expands the target." This is a bare account of an incredibly complex system, but we can see at least some critical aspects of this knowledge system that separated it from the navigation of Cook's voyage: its attachment to orality; its consideration of the canoe as a place of stasis, around which islands "move"; its understanding of navigation as linked to what anthropologists have called "performance" or "actions," rather than to calculations.[60]

Tupaia's magisterial combination of these forms of knowledge looks even more impressive on the latest detailed account from Lars Eckstein and Anja Schwarz, who give us the following scenario: a European hand (probably Parkinson's) begins the map with a Mercator-style projection of the Society Islands, through which Tupaia, onboard the *Endeavour*, had just navigated the ship.[61] Tupaia, they then postulate, was asked to add the additional islands from his own knowledge base (which he had already demonstrated to Cook with a list of over seventy additional islands). Here, though, when Tupaia adds to this projected image, he does so based on his own knowledge system, based on the islands' "directionality" and "sequence."[62] Moreover, he retools the space of the map itself, ignoring the compass directions that guide its initial European construction and adding to its center instead the marking "avatea," which serves to "override the absolute cardinal orientation prefigured for him."[63] The effect is to "quit the abstract cartographic space" and make each island "a centre in its own right, from which a pahi could depart on a specific traditional voyaging path."[64]

Banks does not tell us what he learned from Tupaia, but we know that he was immediately present at some of the most staggering of Tupaia's knowledge collaborations. We know from handwriting comparisons that he worked intimately on the first map of Raiatea with Tupaia, transcribing place names, and that he probably worked too on this second, more comprehensive map of the islands, the main copy of which was found in his papers.[65] At the very least, we can say that he had Tupaia's knowledge-making example in front of him, not only a measure of how far cross-cultural thinking could go, but a model for how to think about visual representations in terms of systems that privilege embodied knowledge.

Drawing Society

In thinking about the ceremony as a process of transformation, Banks turns back to the craft roots of natural philosophical experimentation and, at the same time, toward the indigenous craft practice he sees in front of him. Tattooing is a particular fascination of Banks's, and I would like to turn to the kind of representation that he chooses for it in the journal and what it can tell us about his understanding of craft as a bridge between cultures. In this section I take stock of an additional appearance of craft knowledge in the journal in Banks's series of small drawings. In this case, as with Banks's account of ceremony, craft is engaged with a form of knowledge that is explicitly social. This may be what he takes from Tupaia's knowledge experiments and what he sees of tattooing: that the work of the body is not the work of the individual body but that of the community. We have already seen one form of this in the embodied sociality that crops up in the mourning performance. Likewise, when he draws, Banks is keen to consider which aspects of making might connect the self to another across a cultural divide.

Sprinkled throughout Banks's pages of journal writing are fifteen small drawings, depicting what eighteenth-century Britons called "artificial curiosities." At first it is striking that Banks depicted anything at all in visual terms, traveling as he was with professional draftsmen, like Parkinson, who were called in to paint or draw. Yet there is a stark division between Parkinson's detailed illustrations and the rough little sketches Banks creates. Indeed, Banks's drawings look much like rough sketches of natural phenomena one might see in the period, but the journal's sketches are entirely of manmade things, "artificial curiosities" as opposed to "natural" ones.[66] In total, the objects he draws fall into three categories: patterns (of tattoos, of cloth, of ornament on a building); tools (bait, tattooing instrument, cloth beater, fish hooks); and the shape of built structures (canoes, a "godhouse"). Despite his ongoing botanizing Banks does not draw natural specimens in the journal, turning only to this kind of representation when he wishes to depict nature shaped or arranged by the human hand.

The majority—eleven—of Banks's drawings depict objects and patterns he saw in Tahiti, with an additional three devoted to objects in New Zealand. But before the *Endeavour* arrives at Matavi Bay, Banks has already sketched twice. I'd like to begin with his first sketch, which shows Banks working out a way of proceeding and experimenting with the relation between written text and drawing—a process that will become more elaborate in Oceania. Banks's first drawing (fig. 4) is typical of most of the drawings later in the journal: it is small, incorporated into the lines of writing, roughly drawn, and of a made or constructed object. As Banks tells us, it is a "curiosity." Before the ship has left Europe, they touch down in Madeira, just off the

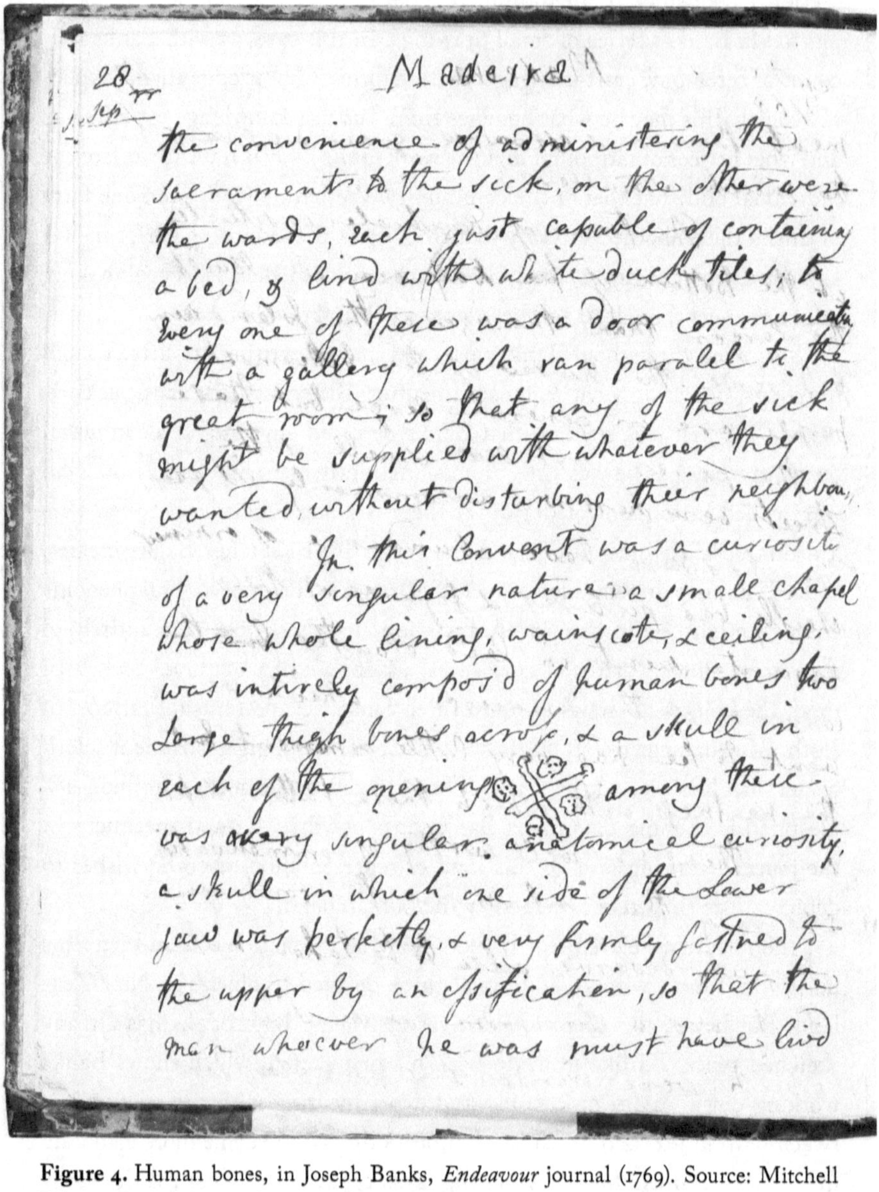

Madeira

the convenience of administering the
sacraments to the sick, on the other were
the wards, each just capable of containing
a bed, & lind with white dutch tiles; to
every one of these was a door communicating
with a gallery which ran paralel to the
great room; so that any of the sick
might be supplied with whatever they
wanted without disturbing their neighbour.

In this Convent was a curiosity
of a very singular nature a small chapel
whose whole lining, wainscote, & ceiling
was intirely compos'd of human bones, two
large thigh bones across, & a skull in
each of the openings; among these
was a very singular anatomical curiosity,
a skull in which one side of the lower
jaw was perfectly, & very firmly fastned to
the upper by an ossification, so that the
man whoever he was must have livd

Figure 4. Human bones, in Joseph Banks, *Endeavour* journal (1769). Source: Mitchell Library, State Library of New South Wales, SAFE/Banks Papers 03.01.

coast of Portugal. Banks turns immediately to observing cultural artifacts of a familiar sort, what he calls church "ornaments": "The Churches here have abundance of ornaments, cheifly bad pictures and figures of their favourite saints in lac'd cloaths" (B 1:163). Following this dismissive estimation, he finds something a good deal more interesting in a Franciscan convent, which he calls a "curiosity of a very singular nature":

> a small chapel whose whole lining, wainscote, and ceiling, was intirely compos'd of human bones, two large thigh bones across, and a skull in each of the openings. Among these was a very singular anatomical curiosity, a skull in which one side of the Lower jaw was perfectly and very firmly fastned to the upper by an ossification, so that the man whoever he was must have livd some time without being able to open his mouth, indeed it was plain on the other side that a hole had been made by beating out his teeth, and in some measure damaging his Jaw bone, by which alone he must have receivd his nourishment. (B 1:164)

Banks proceeds from curiosity to curiosity, each containing the next: he begins with the "curiosity of a very singular nature," the unusual bone-filled chapel; then moves us to the unusual, ossified bone; and, finally we reach the strange modification of the bone itself, the "hole" that has been drilled for eating. Banks the scientist, the trained observer, keeps going closer, motivated by "curiosity," which leads to greater and greater knowledge of the environment. He also ends the passage with the ultimate curiosity: the ossified and further modified bone. Such a bone would have been at home in a cabinet of curiosities, or in a virtuoso's collection of anatomical abnormalities. The prose passage does its work, leading the reader through this investigation and leaving us, scientist or collector, with the object itself.

But the small sketch of skulls and crossbones that Banks draws on the page does a different kind of representational work. The word *illustration* might seem to apply to what we have here, but we should use that term with caution. I think of Barbara Maria Stafford's account of the way that the lavish illustrations in printed travel books of this era complete or fill out prose descriptions.[67] The sketch, however, doesn't seem to fill out, or even follow along. If the passage drives toward the strange bone, the drawing takes us back to the arrangement on the walls. And it has a remarkably different emphasis. Whereas the passage related particularity, detail, the sketch shows relation. Whereas the passage seemed attached to specimen or to an object in a collection, the sketch shows a pattern. And whereas the plain prose of the passage aimed for specificity, for accuracy, the sketch is rough. We can barely make out what Banks is up to here, and we certainly learn nothing new about the appearance of objects from this visual depiction of them.

So far, I have put the sketch's contributions largely in terms of the negative: what the passage shows us that the sketch does not. But what does the sketch show, exactly? What kind of knowledge does it present? To begin with, we might note that the hastiness of the sketch draws our attention as much to how it is made as to what it depicts. Its crude shapes seem to cry out for the kind of distinction the much later anthropologist Jasamin Kashanipour calls "immediate sketching" and "practice" by sharp contrast with "art."[68] After drawing hurriedly on a napkin or in the middle of a project, we probably all have said ourselves some version of what Kashanipour says about forms of ethnographical sketching: it's not a work of art, its value is not aesthetic but instead it is meant to do something practical (directions, perhaps, or relations between parts of a building). In a related vein, Daston and Galison describe the drawings of naturalists in this period as "tools to think with."[69] But Banks's small sketch, like Kashanipour's "immediate sketching," goes well beyond this: it is "a search for connectedness."[70] The sketch performs this in visual terms, its handmade quality pulling it toward the lines of handwriting. This book has its own set of Enlightenment terms for this attempt to get beyond realist representation, as when Cowley's Bacon crushes the grapes or Hogarth's line takes in "action."[71] Likewise, the sketch calls attention to the hand, to the act of making that obscures differences between writing and drawing. Banks follows Hogarth in showing both how making might work in relation to the line—its wobbles a testament to handiwork—and the way that abstraction may entail proximity, and materiality, rather than distance. Here, as too in the "performance" of mourning, Banks pushes representation into being. Michael Taussig puts it this way, channeling Walter Benjamin: mimesis can "get hold of something by means of its likeness."[72]

In so doing, the small drawing remakes the conditions for representation that we ordinarily associate with European depictions of objects from the contact zone. Nicholas Thomas describes the way in which "invented curiosities" were subject to "extreme decontextualization, considered as

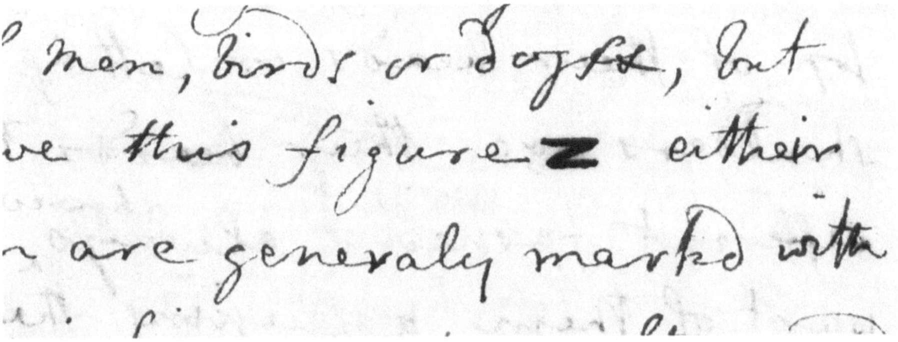

> men, birds or dogs, but
> ve this figure ≈ either
> ~ are generaly marked with

Figure 5. Tattoo pattern, Tahiti, in Joseph Banks, *Endeavour* journal (1769). Source: Mitchell Library, State Library of New South Wales, SAFE/Banks Papers 03.01.

> of that had black marks ab
> under their armpits the sid
> indented ▲▲▲ they had also
> round their arms & legs &
> singular as well as the cloth

Figure 6. Tattoo pattern, Tahiti, in Joseph Banks, *Endeavour* journal (1769). Source: Mitchell Library, State Library of New South Wales, SAFE/Banks Papers 03.01.

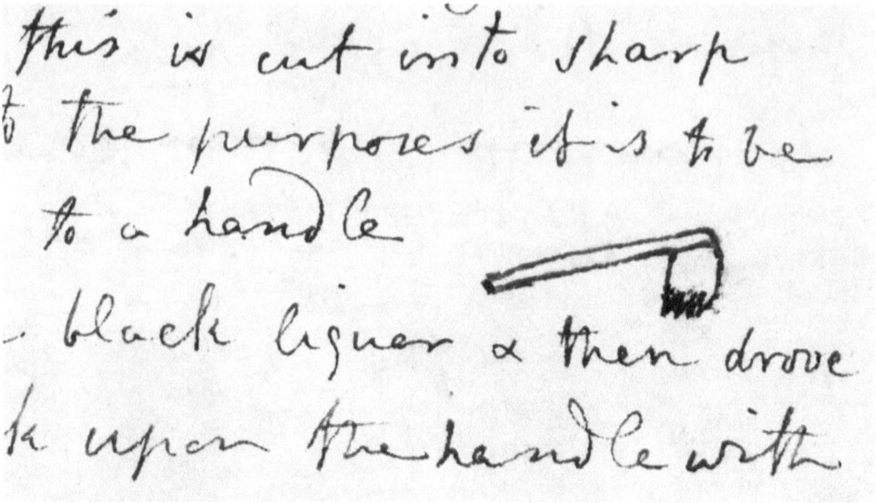

> *this is cut into sharp
> to the purposes it is to be
> to a handle*
>
> *— black liquor & then drove
> it upon the handle with*

Figure 7. Tattooing instrument, Tahiti, in Joseph Banks, *Endeavour* journal (1769). Source: Mitchell Library, State Library of New South Wales, SAFE/Banks Papers 03.01.

specimens."[73] Harriet Guest associates this with "curiosity" in particular, which plucks out and "thrives on the isolation of its exotic object."[74] Banks's drawings work against these impulses, even when his written text embraces them. Instead, the drawings offer the possibility that how a thing is done might connect a representation to its original—and thus to the world of which that original is a part.

We can now turn to Banks's representations of tattoo patterns (figs. 5 and 6) and a tattooing instrument (fig. 7) in the journal. As we already have seen, tattooing is a special kind of representation, involving the "painted" body and a paint that works as a kind of jeweled "inlay," making inside/outside distinctions difficult. In his summary section on the "customs" of the "south seas," Banks turns to tattooing as a kind of recipe, telling us how it is done. His sketches offer something more, along the lines of the pattern of bones. But here tattoo patterns interact directly with the lines of Banks's script. Indeed, the zigzag pattern is like a letter of Banks's own alphabet (he calls it a "figure") and nearly disappears into the line as "Z" (fig. 5). The small triangle pattern, by contrast, is distinct from writing but set inside the line in such a way as to draw our attention to its resemblance to Banks's script (fig. 6); Banks's description of the pattern as "black marks" deepens its connection to his handwriting. In an extraordinary layered attempt at "connection" through representation, Banks merges drawing and writing, even as he also remakes in the fashion of the bones. If Sprat talked about words getting close to things, the closeness Banks has in mind works a bit differently, miming the process of the thing in the world with the representation itself. In his representation of the tattooing instrument (fig. 7), Banks renders it with tiny drops of tattoo ink, as though to touch one kind of ink with another. Like Tim Ingold, Banks wants to remind us here that writing is utterly material and is just another form of drawing, more similar to needlework or tattooing than to the printed map.[75] In writing turned away from print and back to its origin in craft and drawing, we are a long way from Latour's "immutable mobiles."

Within his consideration and reconsideration of representation, Banks provides his most intriguing accounts of what it might mean to access an unfamiliar society. Banks's consideration of knowledge as participatory, and his use of craft to imagine a profoundly social identity (and within terms that he perceives to be that other society's own) are entirely out of step with the way that the history of anthropology—and its relation to literature—is ordinarily told. Rather, Banks's position here is most similar to later "theorists" like Claude Lévi-Strauss or Pierre Bourdieu, who have attacked the problem of cultural knowledge by going back to "experience" itself, and refashioning that experience through craft in the figure of the bricoleur and the concept of habitus. Banks seems to appreciate some part of this much later motivation: that a knowledge of society could press us to examine what kind of "experience" we are using to examine that other set of persons and relations, in the first place.

Hazy Sight

In one of this chapter's first examples of craft brought together with perception, Banks notes that the weather is "very hazey and thick." This is the moment when Tahiti is "so faint that few could see it" (B: 1:249). The historian Paul Carter uses similar language in *The Road to Botany Bay*, where Carter begins by having us think beyond an imperialist history that depends on an "all-seeing spectator."[76] In lieu of this, he presents his own book by gesturing toward a "world of experience" put this way: "*The Road to Botany Bay* is concerned with the haze which preceded clear outlines."[77] His aim is not geography, he tells us, but "the spatial forms and fantasies through which a culture declares its presence."[78]

How is it that the imperialist comes so close to the anti-imperial historian? Carter and Banks bring us back again to questions about craft, representation, and ethics. Well beyond Carter's textual example, we can see the way in which Banks's experiments with representation might seem to work toward ethical ends, broadening "experience," establishing connection with other people and with the world. In this way Banks would seem to work around forms of representation that are associated with imperial authority and dispossession: the all-knowing eye; the realist particular.[79] But Banks shows us how these later understandings of representation do not hold well in this earlier historical moment. Banks thinks and works in both spheres at once, collecting and taxonomizing, even as he is able to see beyond the limitations of that experience. His example cautions us not to embrace one way of seeing or experiencing because—and this seems to be Carter's hope—it might undo the other.

Moreover, even though we are operating in the realm of perception, it is worth revisiting Richard Sennett's caution in *The Craftsman* that craft does not have any necessarily ethical valence, that we should remember that it includes atomic bombs as well as woven baskets.[80] We can see the especially slippery categories of experience at play in another Baconian moment in Banks's journal, this one also concerned with craft. Here, I turn to Banks in Raiatea, on the day after "Tupias boy Tayeto" has explained to him that the "building" he has seen is called "house of the god" (B 1:316). On the first day, Banks finds the "house" empty and looks but does nothing more, noting that he could "offend the people" (B 1:316). But on discovering similar buildings the next day, he cannot resist: "One of these I examind by putting my hand into it: within was a parsel about 5 feet long and one thick wrappd up in matts, these I tore with my fingers till I came to a covering of mat made of platted Cocoa nut fibres which it was impossible to get through . . . especially as what I had already done gave much offence to our new freinds" (B 1:318). Banks reminds us that tactile knowing may be bound up with cultural connection or with cultural violation. Indeed, he seems to enact a kind of primal Baconian scene here, inserting his body into the sacred object, as though to determine its meaning, or "use." Bacon's metaphors for knowing, after all, are markedly violent. In his account, the knower, the craftsman, doesn't just come to be aware of the powers of nature; he brings those forces of nature to heel.[81] In this passage, experimental knowledge meets indigenous craft, the hand hits a well-woven mat in an effort to break through, and the knowledge of making veers into destruction and desecration.

FIVE

Use

USELESS BODIES IN JOHNSON AND BOSWELL

IN *THE JOURNAL OF A TOUR TO THE HEBRIDES,* the published record of his 1773 trip to Scotland, James Boswell tells a story about his famous friend Samuel Johnson meeting the Scottish Lord Monboddo. Boswell sets the scene for us: Monboddo greets the visitors in a "rustick suit, and wore a little round hat."[1] In this garb the lord tells his visitors that they "now saw him as *Farmer Burnet*" and that they shall be treated to a "farmer's dinner" (JB 198). It turns out that Monboddo has a very particular farmer in mind:

> He produced a very long stalk of corn, as a specimen of his crop, and said, "You see here the *laetas segetes:*" he added, that *Virgil* seemed to be as enthusiastick a farmer as he, and was certainly a practical one.—*Johnson.* "It does not always follow, my lord, that a man who has written a good poem on an art, has practised it. Philip Miller told me, that in Philips's CYDER, a poem, all the precepts were just, and indeed better than in books written for the purpose of instructing; yet Philips had never made cyder." (JB 78)

The situation may be unusual: the famous Englishman on tour meets with a Scottish lord and philosopher in farmer's dress, speaking about his corn crop in the Latin of Virgil's *Georgics*.[2] But the epistemological stakes of this exchange and even its terms are already familiar from this book's discussions of craft knowledge. Here, at the outset of *Journey,* Boswell gives us Samuel Johnson as the representative of theory—indeed, of practice-as-theory. In his description of the trip Boswell has us associate Johnson with technical knowledge related to making—he mentions Johnson's conversational displays of expertise on tanning, coining, and brewing—even as Johnson remains always the "abstract scholar" and "philosopher" (JB 278, 344). As Boswell paints him here, Johnson is the embodiment of a relatively new form of intellectual laborer, taking his cue from the georgic, where mental labor (as we saw Duck and Collier contemplate it) was thought in terms of and, at the same time, removed from the scene of work.

This chapter aims to make differently meaningful some of Boswell's apparently ordinary terms for Johnson. After all, what is more obvious than to refer to Samuel Johnson as a "sage" or a "philosopher."[3] Who could better represent the labor of knowledge—vast and immersive, yet also distanced—than the man nicknamed, after the most famous of his many significant literary endeavors, Dictionary Johnson? It is the burden of this chapter to show that Boswell's positioning of Johnson as a particular kind of "abstract" or removed thinker is not mere presentation of matter of fact but a deliberate form of argument. Moreover, what *philosophy* means is much less certain than Boswell's characterizations of the great man seem at first to suggest. In both Johnson's and Boswell's accounts, the domain of practical use and the body's relation to thought and experience will prove central; what results, I argue, is not a single form of embodied cognition but a set of questions that requires us variously to reconceive thought and action in relation to the body.

The commonplace critical account of Johnson's and Boswell's texts is that "the main focus in Johnson's account is Scotland, in Boswell's it is Johnson himself."[4] But I would like to suggest this does not sum up the two published books, so much as it offers us an opening to consider the strangeness of their differences. Over twelve weeks the sixty-three-year-old Johnson, embodiment of Englishness, traveled with his friend, the thirty-two-year-old Boswell, whose Scottishness has been described as a "deeply perplexed national self-understanding."[5] The two men shared experiences both immediately and in the accounts of the trip they wrote and exchanged. As John B. Radner puts it, the two men "collaborated and competed in narrating the trip."[6] We know that during those weeks in the fall of 1773 Boswell wrote—in very refined prose—his account as the journey proceeded and that Johnson read passages of the journal in Scotland. Johnson recorded the trip in a notebook (now lost), and he shared his relations with Boswell. Once the trip was complete Johnson borrowed Boswell's journal when he composed his own text for its

1775 publication. We know that Boswell had hoped to publish his own text earlier but held off until Johnson's death;[7] he and Edward Malone revised Boswell's journal and the published *Tour* appeared in 1785. This makes for a web of influences, rather than any unidirectional vector, and we can see both men working through related ideas in closely related terms. Moreover, although the "philosophical" aspects of writing are almost always discussed in Johnson's narrative, I will show ways in which Boswell commandingly sets philosophical terms, both for what appears in Johnson's account and in his own published response to it.[8]

The goal of this chapter is to illuminate the ways in which theory and practice, those distinct ways of knowing the world, come to be meaningful for these accounts of what Johnson calls "the manners of a people" (SJ 22).[9] Johnson is often quoted as saying that "Our business was with life and manners" (SJ 32). Johnson's and Boswell's trip began just as the *Endeavour* voyage was published in June in Hawkesworth's *Voyages*, and Rogers notes that "no literary work of any description was more vibrantly alive in the public consciousness when Johnson took a coach for Edinburgh on 6 August."[10] Unlike that South Sea voyage, of course, the trip north to Scotland does not begin with major scientific projects: there is no transit of Venus to record, no plant specimens to collect. From the outset the two men see this as a social and historical venture. This does not mean, however, that Royal Society standards for knowledge production are far from the travelers' minds. Johnson was steeped in the major seventeenth-century texts of natural philosophy; as Richard Schwartz has remarked, Johnson's reading for the *Dictionary* alone included the texts of Francis Bacon, Herman Boerhaave, Robert Boyle, Isaac Newton, John Wilkins, and John Woodward.[11] Johnson and Boswell both would have had in mind the way this natural philosophical context emerged in travel writing on Scotland in particular. There are two texts that directly inform the men's own reports on Scotland: Martin Martin's *Description of the Western Islands of Scotland* (1703), a favorite of Johnson's as a boy, and a copy of which was brought on the 1773 trip; and Thomas Pennant's *A Tour in Scotland* (1771), published just two years before the trip (JB 13, 157). Although quite different, both of these texts are written by writers who identify themselves explicitly with the Royal Society and its tradition of knowledge. Martin explicitly refers to the "Natural and Experimental Philosophy" that will guide his approach, one which will "study Things there more than Words."[12] Pennant, both natural historian and a fellow of the Society of Antiquaries, was also the author of a text on zoology, and he published an account of an earthquake in the Royal Society's *Philosophical Transactions*.[13] Johnson refers to the great accuracy of Pennant's approach, and Boswell grabs a copy of Pennant off a shelf at Fort George to study his description of the fort's arrangement (JB 124). These earlier models are very much in mind, then, but also present physically during the trip, as well as during later moments of composition and editing.

As we know from Cook and Banks, however, it was hardly obvious what it might mean to take "things" over "words" or an "experimental" approach to an unfamiliar society. Questions about how society should be conceived or reported on also filter into the journey from an intellectual milieu that is manifested physically on the trip: the Scottish Enlightenment world of Edinburgh. Before Johnson and Boswell reach Monboddo's estate, they have already encountered some of the most prominent philosophers of the Scottish Enlightenment, all of whom are at work, in one way or another, on questions surrounding the origins of civil society, the history of Scotland and other "rude" societies, and even the nature of "experience." In Edinburgh Johnson meets William Robertson (author of *The History of Scotland*, 1759),[14] Adam Ferguson (author of *An Essay on the History of Civil Society*, 1767), Hugh Blair (literary critic and author of an extensive defense of Macpherson's Ossian poems); in Glasgow, he meets Thomas Reid (author of *Inquiry into the Human Mind, on the Principles of Common Sense*, 1764) (JB 32, 42, 394, 369). Johnson and Boswell dine and converse, then, with men at the vanguard of the country's intellectual movement whose ideas inform their immediate analysis of Scotland. Boswell mentions Robertson's and Ferguson's texts by name to remind the reader of this fact, and Ossian is a topic of conversation across the entire journey, both in these Lowland intellectual circles as well as once the visitors reach the Highlands (JB 178, 177).

We can return to Monboddo to consider more fully just what it meant to study "the manners of a people" in Scotland in 1773, for just that single anecdote unfolds for us the kind of terms with which Johnson and Boswell had to wrestle. Monboddo has just published the first volume of his major work, *Of the Origin and Progress of Language*, a consideration of language and, more broadly, the natural history of mankind. During the visit to the estate Boswell begins joking about Monboddo's "primitivism" (as we might now call it) right away: Monboddo, gesturing toward the Douglas arms on his house, introduces his lineage by saying that "our ancestors . . . were better men than we" (JB 77). When Johnson replies with the opposite—that the modern man is better—Boswell worries that "This was an assault upon one of Lord Monboddo's capital dogmas" (JB 77). Before the men have even entered the house, then, Boswell has us frame their meeting in terms of an understanding of humanity's rise or fall, progress or degeneracy. Monboddo's most famous hypothesis—that men were descended from apes—is wildly mocked by Johnson at several moments during the *Journey*. But here at the house, as later, the joking nature of the response does not at all diminish the extent to which the observation of "the manners of a people" is wrapped up with the definition of the history of humanity.

A central example of Scottish Enlightenment thinking in the *Journey* involves Johnson's use of a stadial theory of society (originating in Smith but animating many Scottish conjectural accounts) to make an argument about the status of Scotland's "useful" arts. This is, to begin with, an excuse for Johnson both to situate Scotland at an earlier stage of social development and to marshal England's superior know-how as a kind of correction or help to the Scots. But Johnson's account of what is "useful" and what "useless" ends up going deep and indeed plunges him back into a justification for the direct experience of the traveler—that is, for the grounds of his own knowledge project. Bacon had supposedly answered this point a century earlier, in part by turning to craft or maker's knowledge. Johnson returns to the origin of this resolution and plays out both its immediacy and its insufficiency in its apparent disregard for the body's involvement in experience. Boswell, for his part, engages ideas of use in his own text, centering them on Johnson's useless body and employing them to frame his own form of practical knowledge.

The Uselessness of Scotland

In 1755 William Robertson, the divine and principal of Edinburgh University who dined with Johnson multiple times on the journey, said in a sermon that in the Scottish Highlands, "society still appears in its rudest and most imperfect form."[15] Although in this period America was by far the most common case of this "rudeness" (in the travel writings of Pierre-François-Xavier de Charlevoix and Joseph-François Lafitau, for example), the cutting-edge historians of humanity in Glasgow and Edinburgh could point to the northern reaches of their own country for an example of an earlier civilization.[16] Twenty years later, however, Johnson and Boswell find only remnants of this "imperfect form," greeted instead by a country modernized and colonized. As Johnson puts it, "There was perhaps never any change of national manners so quick, so great, and so general, as that which has operated in the Highlands, by the last conquest, and the subsequent laws. We came thither too late to see what we expected, a people of peculiar appearance, and a system of antiquated life" (SJ 57). In the wake of the suppression of Highland life after the failed 1745 rebellion, Johnson and Boswell find dress and weapons regulated, the feudal structure on the point of collapse, military roads carved through the landscape.

In order to assess just where Scotland is in its development, Johnson employs a framework developed by Scottish Enlightenment thinkers: the stadial theory of society laid out by Smith that traces progress in terms of modes of subsistence, from the age of hunters to the age of shepherds, to the ages of agriculture and commerce.[17] Sometimes Johnson, traveling through the Highlands, can be found sorting these subsistence-based ages of development—or trying to—as he looks at the individual instances in front of him. Take his first encounter with the family in a Highland hut:

> When we entered, we found an old woman boiling goats-flesh in a kettle. She spoke little English, but we had interpreters at hand; and she was willing enough to display her whole system of economy. She has five children, of which none are yet gone from her. The eldest, a boy of thirteen, and her husband, who is eighty years old, were at work in the wood. Her next two sons were gone to Inverness to buy "meal," by which oatmeal is always meant. Meal she considered as expensive food, and told us, that in spring, when the goats gave milk, the children could live without it. She is mistress of sixty goats, and I saw many kids in an enclosure at the end of her house. She had also some poultry. By the lake we saw a potatoe-garden, and a small spot of ground on which stood four shucks, containing each twelve sheaves of barley. She has all this from the labour of their own hands, and for what is necessary to be bought, her kids and her chickens are sent to market. (SJ 33).

Johnson's details allow the reader to track the kind of "subsistence" he sees in Loch Ness. This woman's family remains chiefly dependent on animals. Although it is possible to see commerce around the edges of this life—the two boys at market are there to buy oatmeal—Johnson lets us know that the family is only partly in this arena, retreating into a mode of subsistence with only what they have by "their own hands," through goat milk and gardening. We don't see proper agriculture here (just a small number of potatoes and barley, which could be ranked with the incidental corn that Smith mentions for the indigenous North Americans); indeed, Johnson refers to the woman's welcome as "pastoral hospitality" (SJ 33). Elsewhere, along the same lines, we find Johnson offering similar salient details—whether shoes are made within the domestic space or as a commercial product; whether or not plows are used in farming—that allow us to understand Scotland's proximity to a modern commercial society.

Johnson wrestles, then, with the unevenness of the stadial model when it comes to assessing a colonized society forced to modernize. Within this account of progress, however, Johnson finds Scotland to be an exception to economic progress on a key front. In what will become a near refrain over the course of the *Journey*, Johnson early declares Scotland deficient in the "useful" or "manual arts." As he explains it, "I know not whether it be not peculiar to the Scots to have attained the liberal, without the manual arts, to have excelled in ornamental knowledge, and to have wanted not only the elegancies, but the conveniencies of common life" (SJ 28). Here, for the sake of argument, perhaps, Johnson seems to grant the connection established by the supporters of Ossian: as Hugh Blair puts it, the "artless ages" are those of great, passionate poetry.[18] For Johnson, looking at Scottish society, the problem is that this has not helped a situation more dire: the more fundamental "arts," those that could ground commerce and thus modernity, never came. These men of ornament and imagination "were content to live in total ignorance of the trades by which human wants are supplied, and to supply them by the grossest means" (SJ 28).

Johnson's account of a useless and ignorant Scotland paves the way for an account of colonial power as useful. Johnson shores up his point about useful arts in his extensive focus on agricultural progress and method over the course of the book, a focus informed by John Locke's influential claim in the *Second Treatise of Civil Government* (1689) that the ownership of land depends upon its proper cultivation and use. Referring to the Union of 1707, Johnson argues, "the culture of their lands was unskillful" until "the Union made them acquainted with English manners" (SJ 28). But the Union was not enough.

Johnson finds lack of skill and lack of use wherever he goes. Early in the *Journey*, still in the Lowlands, he comments that although Cromwell's soldiers taught the inhabitants to plant kale, they "cultivate hardly any other plant for common tables" (SJ 28). Once he reaches the Highlands, in Col, after stressing Young Col's introduction of "the culture of turnips," Johnson encourages more: "Wherever heath will grow, there is reason to think something better may draw nourishment; and by trying the production of other places, plants will be found suitable to every soil" (SJ 124–25). Johnson's famous obsession with the absence of trees in Scotland begins early, near Bamff, with the declaration that he had "seen only one tree not younger than myself" (SJ 21). This barrenness of landscape is revisited many times over the course of the trip, culminating in an extensive discussion of tree planting—in relation to georgic principles—in the Highlands and an examination of Sir James Macdonald's (unsuccessful) attempt to plant trees "in part of the wastes of his territory" (SJ 140).

In his project to deprive Scotland of the useful arts, Johnson also omits the arts that earlier travelers display in their accounts of the country. This is particularly true in relation to Martin, whose account is focused on native medicine ("their Admirable and Expeditious way of Curing most Diseases by Simples of their own Product," as he puts it on the title page). After all, manual arts do not need to wait for commercial society in order to reveal skill and mastery. We recall Banks's account of weaving and his drawings of woven cloth in Tahiti. Martin's description of Scotland, for its part, describes extensively the weaving of plaid cloth in the Highlands. He relates that the "*Plad* wore only by the Men, is made of fine Wool, the Thread as fine as can be made of that kind."[19] His description covers the "ingenuity requir'd in sorting the Colours," and the patterns women make "upon a piece of Wood, having the number of every thread of the stripe on it."[20] And plaid is useful: Martin tells the reader how the plaid is worn on the body and comments that "this Dress for Footmen is found much easier and lighter than *Breeches*, or *Trowis*."[21] Not only does Johnson omit any account of the making of plaid; when he mentions the garment, he diminishes it on these very grounds of "use." If he concedes that the Scots have made something, he is sure to point out that it is not useful. Plaid for him is "incommodious and cumbersome dress," and, on his account, its main good use is for sleeping (in a pinch): "they could commodiously wrap themselves in it, when they were obliged to sleep without a better cover" (SJ 52).

We can see Johnson's attempt to extract skill from making in his account of Anoch. Of the house where they stay Johnson observes that "the part in which we dined and slept was lined with turf and wattled with twigs, which kept the earth from falling" (SJ 36). Boswell's take on this method of building is different: "The house here was built of thick turfs, and thatched with thinner turfs and heath. It had three rooms in length, and a little room which projected. Where we sat, the side-walls were *wainscotted*, as Dr. Johnson said, with wicker, very neatly plaited. Our landlord had made the whole with his own hands" (JB 136; emphasis original). What comes across in Johnson as a mere dirt house is in Boswell's account a description of sophisticated craftwork. We notice the way in which Johnson has taken out the comment Boswell ascribes to him about "wainscoting" and Boswell's own account of plaiting—descriptions that highlight the precision of the handiwork (and indeed connect it to other crafts, such as carving or basket weaving) and stress its ornamental quality in addition to its practical one of holding up the wall. We see in Boswell's account, too, a stress that M'Queen has "made the whole with his own hands"; Johnson seems less concerned with how this was made or by whom, whereas in Boswell the house is the product of a particular craftsman.

Let's return, though, to that uncomfortable plaid. For when Johnson's accounts of the "useful arts" appear in his descriptions of Scotland, they regularly end up as this one does: in technological failure. As with that "incommodious" plaid of the Highlands, so with Scottish windows Johnson describes in Bamff, still in the Lowlands:

> The art of joining squares of glass with lead is little used in Scotland, and in some places is totally forgotten. The frames of their windows are all of wood. They are more frugal of their glass than the English, and will often, in houses not otherwise mean, compose a square of two pieces, not joining like cracked glass, but with one edge laid perhaps half an inch over the other. Their windows do not move upon hinges, but are pushed up and drawn down in grooves, yet they are seldom accommodated with weights and pullies. He that would have his window open must hold it with his hand, unless what may be sometimes found among good contrivers, there be a nail which he may stick into a hole, to keep it from falling. (SJ 21–22)

And, as with windows, so it is with "brogues," the Scottish shoes that Martin described as "anciently . . . a piece of the Hide of a Deer, Cow, or Horse, with the Hair on, being tied behind and before with a Point of Leather."²² No such neutral description is possible for Johnson, who weighs in on quality of the material and design of what he pointedly calls "artless shoes":

> In Sky I first observed the use of brogues, a kind of artless shoes, stitched with thongs so loosely that though they defend the foot from stones, they do not exclude water. Brogues were formerly made of raw hides, with the hair inwards, and such are perhaps still used in rude and remote parts; but they are said not to last above two days. Where life is somewhat improved, they are now made of leather tanned with oak bark, as in other places, or with the bark of birch, or roots of tormentil, a substance recommended in defect of bark, about forty years ago, to the Irish tanners, by one to whom the parliament of that kingdom voted a reward. The leather of Sky is not completely penetrated by vegetable matter, and therefore cannot be very durable. (SJ 50)

In the passages on windows and shoes, Johnson criticizes Scottish methods, contrasting them to superior English methods of building, tanning, and design. Speaking from a society of greater technological advancement, Johnson consistently marks Scottish objects in terms of the functional failure that results from them: the bulky loose plaid that one must hold to the body (it "require[s] one of the hands to keep it close"); the window that won't stay open; the leaky and short-lived shoe (SJ 52). At every turn the English and the Englishman signal the useful, the Scottish the useless. In the examples from agriculture, barrenness for Johnson signals both the lack of the "art" that would result in proper agriculture and ties back into the grounds for colonizing Scotland in the first place. Technological failure has a related valence. As Sara Ahmed puts it, "the failure of things to work creates an incentive to make new things."²³ We can see in Johnson's example that the use of land is not the only way that colonial power justifies itself in Scotland; the need to craft better, more technologically advanced, objects demands the knowledge of an outsider.

Johnson's claims about windows and shoes are straightforwardly committed to the message that English technology is better and that the Scots have a lot to learn.²⁴ But Johnson suggests that they have another purpose too. His passage on windows is followed by a reflection on the kind of description he has just offered:

> These diminutive observations seem to take away something from the dignity of writing, and therefore are never communicated but with hesitation, and a little fear of abasement and contempt. But it must be remembered, that life consists not of a series of illustrious actions, or elegant enjoyments; the greater part of our time passes in compliance with necessities, in the performance of daily duties, in the removal of small inconveniencies, in the procurement of petty pleasures; and we are well or ill at ease, as the main stream of life glides on smoothly, or is ruffled by small obstacles and frequent interruption. The true state of every nation is the state of common life. The manners of a people are not to be found in the schools of learning, or the palaces of greatness. (SJ 22)

Johnson quickly flies from the mention of embarrassingly "diminutive observations" to the general language of the moral philosopher. Only a tiny moment of "abasement" has occurred in this explanation designed to reassure the reader. But we should also look back at just what this "common life" consists of: an intimate knowledge of window panes and the technological failure of the Scottish window. Critics have read this passage as referring to the connection of low and common detail with "common life."²⁵ This is a connection made elsewhere in prose of the century, of course, in its "realism," for which Johnson might seem to make another bid here. But the passage Johnson pinpoints for apology is not "common" in its detail. It is committed, rather, to description of function, that purpose that lies, as it were, within the object itself. Without giving us much clue as to why this would be the case, Johnson associates what Gilbert Ryle calls "knowing how" (as opposed to "knowing that") with the study of "manners" and of "common life." Ryle describes "knowing how" as bringing together "the allegedly incompatible properties of being *kith* to theory and *kin* to practice."²⁶ For him this is intelligence in performance, a knowledge "actualised or exercised in what he does"; it is a knowledge "in practice."²⁷

As with Defoe's craft realism, Johnson's conflation of "diminutive observations" and how-to knowledge should give us pause as to what we are looking for in his detailed prose. More important, though, is the suggestion—albeit a bare one here—that there is something about society that requires a knowledge tending toward the practical. Johnson's colonial how-to knowledge is a crude starting point for this notion, but Johnson and Boswell both develop other ways of comprehending the basic premise that to know Scotland requires a specialized form of knowledge not easily reducible to "fact" or "experience."

Useful Aesthetics

In *A Journey to the Western Islands of Scotland*, this "knowing how" is colonial, a knowledge that barrels over local practices, attempting to resolve them into Reason and modernity. It is also, it turns out, a truly unstable way of discriminating between English and Scottish, between Johnson and those he visits. Johnson himself attends to the instability of "use" as such a marker when he turns back the "useful" on his own knowledge project.

Immediately before this turn, Johnson makes a declaration that should hardly surprise us: along with the shoes and the windows, the landscape of Scotland is "useless." Looking at hills covered in black heath, he concludes:

> What is not heath is nakedness, a little diversified now and then by a stream rushing down the steep. An eye accustomed to flowery pastures and waving harvests is astonished and repelled by this wide extent of hopeless sterility. The appearance is that of matter incapable of form or usefulness, dismissed by nature from her care and disinherited of her favours, left in its original elemental state, or quickened only with one sullen power of useless vegetation. (SJ 39–40)

We notice that Johnson is still thinking here about cultivation and about the extent to which the land could be better employed to grow food. But his terms are those of eighteenth-century aesthetics; what concerns him here is less what happens with the advancement of society and more what happens with the "eye." It is the eye that seems stymied by the "matter incapable of form or usefulness" that is covered only with "useless vegetation."

This damning conclusion immediately finds itself in the ear of an imaginary interlocutor, who proceeds to turn the problem of "uselessness" back on the writer himself:

> It will very readily occur, that this uniformity of barrenness can afford very little amusement to the traveller: that it is easy to sit at home and conceive rocks and heath, and waterfalls; and that these journeys are useless labours, which neither impregnate the imagination, nor enlarge the understanding. It is true that of far the greater part of things, we must content ourselves with such knowledge as description may exhibit,

> or analogy supply; but it is true likewise, that these ideas are always incomplete, and that at least, till we have compared them with realities, we do not know them to be just. As we see more, we become possessed of more certainties, and consequently gain more principles of reasoning, and found a wider basis of analogy. (SJ 40)

As though it is contagious, the "useless" "matter" declared by Scotland's barren landscape seems to infect Johnson's own project with the status of "useless labours," calling into question its most basic premise in direct experience. "It will very readily occur," Johnson begins. But how readily, and why? By what logic does the "matter" of the Scottish landscape lead directly to the question of whether one can write a travel book by reading other travel books, or whether one must travel through the landscape oneself? Johnson frames the question in terms of aesthetics, but he answers it in terms of Baconian induction, the gradual progress toward "certainties." What connects these two registers is craft: movement from the "matter" that cannot be shaped to the craft-based Baconian perception, that accounts for experience by way of maker's knowledge.[28]

The fresh way that Johnson reconsiders these (by 1773) old terms for experience—offering Baconian natural philosophy as though a new answer to the Scotland problem—should encourage us to look further at what he is trying to pull from both aesthetic discourse and natural philosophy in order to account for the pressure Scotland has put on "experience." We can see further into this matter when it becomes clear that Johnson has crafted much of this passage as a direct response to Joseph Addison's "Pleasures of the Imagination" essays in *The Spectator*. In presenting the "appearance" of Scotland in a description of the mountainous region outside Glensheals, what he refers to as "the bosom of the Highlands," Johnson frames the landscape in Addison's terms for the spectator's response to the landscape (SJ 38). Johnson worries about the "flowery pastures and waving harvests" that are not to be found, and he focuses on the "rivers and rivulets" that are chief examples of the "novel" in Addison's second essay (SJ 39). The passage on formless matter seems designed to be read in these terms. Here is Johnson:

> What is not heath is nakedness, a little diversified now and then by a stream rushing down the steep. An eye accustomed to flowery pastures and waving harvests is astonished and repelled by this wide extent of hopeless sterility. The appearance is that of matter incapable of form or

usefulness, dismissed by nature from her care and disinherited of her favours, left in its original elemental state, or quickened only with one sullen power of useless vegetation. (SJ 39–40)

We see here the thirst for diversity that was important to Addison and the "stream" that almost provides it. We also see the language of "astonishment" as a response to "nakedness" and "wide extent": a clearly-coded "great" (Addison) or "sublime" (Burke).[29] In these terms, we must read Johnson's next account of his own "useless labours" as a way of foreclosing the response that a pleasure-driven aesthetics would entail for the reader. Johnson insists that the "uniformity of barrenness can afford very little amusement to the traveller" (SJ 40). The Baconian induction that comes next, then, takes the place of such "amusement."

If Addison's beauty is known through pleasure, Johnson evacuates the pleasure piece, holding on to knowledge. To use Paul Guyer's formulation, Johnson returns Addisonian pleasure seeking to the cognitive-based aesthetics that Addison—in the history of philosophy—could be understood to have surpassed.[30] This is, to be sure, a strange move. Johnson's target seems no longer the Scotland that can't live up to aesthetic expectations as it does aesthetic philosophy itself which points an account of experience in the wrong direction. Here Johnson follows Abraham Cowley's playbook in the ode "To the Royal Society," in which he galvanizes the "realities" of Baconian experience by comparing them to representations (there, the trompe l'oeil grapes).[31] Johnson, like Cowley, links his mental activity to physical shaping: after all, this Baconian experience is an answer to the question of what to do with Scotland's "matter" that can't be formed. Craft it, Johnson replies in his own analogy, allowing us to see the work of the mind—cognitive, not merely sensual—as the way toward shaping form.

Immediately after his account of a purpose found in "realities" and "certainties," we find Johnson in the aesthetic register, still. And we find him turning to pleasure. The travelers take a break from the ride that day, and Johnson describes them entering a "narrow valley" that was "sufficiently verdant" (SJ 40). In this green space, he says,

> I sat down on a bank, such as a writer of romance might have delighted to feign. I had indeed no trees to whisper over my head, but a clear rivulet streamed at my feet. The day was calm, the air soft, and all was rudeness, silence, and solitude. Before me, and on either side, were high hills, which by hindering the eye from ranging, forced the mind to find

entertainment for itself. Whether I spent the hour well I know not: for here I conceived the thought of this narration. (SJ 40)

From the representational quandary posed by books full of waterfalls and mountains, Johnson takes us back into the world of representation. Indeed, he takes us into the most obviously "artificial" (in the eyes of some eighteenth-century readers, at least) classical mode: the pastoral. Johnson imagines both a "writer of romance" and a moment of invention for that hypothetical writer as a way into the genre. The imagined writer imagines further, and certainties seem already to have dissipated.

Even without this double remove, we can see that Johnson offers no simple embrace of Arcadia. Bradford Q. Boyd associates this passage with Johnson's pastoral allusions elsewhere in the *Journey* and in the Latin poems he wrote during the trip, arguing that across this writing Johnson shows us pastoral scenes "gravely ironized like their ancient models"; Johnson, Boyd show us, digs into the Virgilian pastoral that preceded the empty Arcadian visions of the Renaissance pastoral, and he does so, with a "skepticism of liberal individualism and its sentimental idealizing."[32] In Scotland this means that pastoral works for Johnson as it did for Virgil, allowing him to show two different positions at once: a landscape "improved" and one gutted by invasion, "both the losses and gains of the post-'45 dispensation."[33]

Boyd lets us grasp how Johnson might use the pastoral to encode—and thus to produce—historical meaning in the *Journey*. But Johnson does something else here with pastoral: he uses it. Pastoral, after all, is a literary mode and also a way of seeing the world. Paul Alpers describes Virgil's Meliboeus's representation of "pastoral well-being" as checking the following boxes: "lying in a green spot; seeing a far off (*procul*) sight which both bounds one's world and gives play to the imagination; and, finally, the details and pleasures of innocent feeding."[34] Pastoral is a way of seeing the world in part because it is associated with particular Virgilian characters whose visions do not line up: this pastoral "well-being" is Meliboeus's pastoral, not Tityrus's. But it is also a way of seeing the world because perception is part of its account of place, and its vision is embodied. You see a certain way when you are seated on a verdant bank, watching sheep graze. This complete package is easy for Johnson to import when he finds the (apparent) single spot of grass in an otherwise desolate landscape.

The irony of going from the description of the horrible desolate landscape on one page to the pastoral just paragraphs later is one Johnson wishes for us to grasp. But this ironizing gesture goes beyond the topical complexity that Boyd points out, and returns us to the philosophical questions about experience that Johnson has just asked and (supposedly) answered with Bacon. If we were to ask, "How is Johnson using pastoral here?" we would be on the track that this passage sets out for us. Take Johnson's description immediately preceding that of the romance writer:

> As the day advanced towards noon, we entered a narrow valley not very flowery, but sufficiently verdant. Our guides told us, that the horses could not travel all day, without rest or meat, and intreated us to stop here, because no grass would be found in any other place. The request was reasonable and the argument cogent. We therefore willingly dismounted and diverted ourselves as the place gave us opportunity. (SJ 40)

Johnson carefully situates his deployment of pastoral. He gives its context. He reveals its causes, which are more mundane—and less Arcadian—that we might expect them to be: hunger and thirst, rational instruction and agreement. In the forward-moving logic of the narrative, pastoral is a practical stop, a way station, its ideals firmly buttressed by practical needs.

But I think we would be wrong to assume that what Johnson points to here is simply contrast: between the real world of the journey and the imagined world of the romance writer, between realism and idyllic vision.[35] Rather, Johnson allows us to see the way that both of these worlds involve "use." Elizabeth Fowler's account of poetry and "use" is helpful here. She highlights the way poetry can move us through a "ductile space" and attend to the experience of that movement.[36] Travel writing barely needs such explanation, so at home is it in recording and sharing with the reader the movements of bodies through space: we went here, then there, then there. But pastoral does need this elaboration, and we can see Johnson give it, showing us the way that the pastoral genre intrudes here in order to orient mind and body both: showing how to feel and how to look and imagine. We might say that both the *Journey* and the pastoral example serve as an "aesthetic program[] of bodily experience."[37]

This moves into the next passage, where Johnson comes to his most explicit consideration of aesthetic philosophical concepts developed by Addison and Burke. When Johnson gets, next, to what seems like a full embrace of the terror of Burkean sublimity, in his account of the "phantoms" of "want, and misery" that populate this space in the imagination, he cannot shake this stress on bodily need (SJ 41). There is no sublimity as a movement away from cognition—or, indeed, away from the body.[38] In his third response to the question of how the mind should take in Scotland, he shows us that there is no transcendence even in the world of the phantom. Instead, he insists on a response that punctures "a flattering notion of self-sufficiency" (SJ 41). Here, that looks like bodily collapse: "Whoever had been in the place where I then sat, unprovided with provisions and ignorant of the country, might, at least before the roads were made, have wandered among the rocks, till he had perished with hardship, before he could have found either food or shelter" (SJ 41). Johnson's place in the physical landscape—"where I then sat"—conjures only "imaginations" of bodily dissolution. Above all, what floats "self-sufficiency" and renders thought productive—or, indeed, destroys that self—relies on bodily placement. It depends on where you sit in the landscape.

It's worth thinking about how Johnson moves from Baconian induction to these other forms of literary and aesthetic experience. He presents Baconian induction but with those Addisonian questions naggingly present in the background: How does it make the body feel? Can the landscape produce pleasure? Where do you have to be standing, and how fast does the water have to be moving, for you to feel this? Holding those questions at bay, even for a moment, allows us to see the distance between the spectator they foreground and Bacon's natural philosopher. The effect is to draw our attention to another aspect of how knowledge is made through experience, one that involves how the knowledge-making body is situated in its environment.

Useful Science

In the passages on aesthetics I have just discussed, Johnson swaps what he takes to be Addison's easy answer of "pleasure" for a focus on cognition, but he maintains the focus on the body and its relation to the construction of the world around him. This remains consistent across the text with other moments that are not explicitly aesthetic ones and which in fact pointedly tap into the world of natural philosophy. Natural philosophers in this period have often been read as moving toward nineteenth-century objectivity, privileging a stance of "disinterestedness" or eliminating the body's sensible distractions through the use of instrumentation. For Joanna Picciotto, the Royal Society set the tone for the period following its foundation by keeping the body innocent through new instruments like the telescope and microscope, which could extend the senses beyond their usual, fallen powers.[39] It is in this context that we should read Johnson's own scientific experiments in Scotland.

The first sections of the *Journey*, especially those in cities, seem to place Johnson's text in the tradition of the antiquarian itinerary. Indeed so much has Johnson seemed to work in this vein that critics have quarreled over the degree of his indebtedness to the antiquarian and natural historian Thomas Pennant, whose first *A Tour in Scotland* was published in 1771, before Johnson's trip. Johnson acknowledges drawing on Pennant's account, and at times he seems to have been inspired by the kind of accuracy Pennant's antiquarian itinerary could manage, whether in its calculation of the number of miles between towns, or in its focus on the inscriptions and remains of physical structures that are the stereotypical domain of the antiquary. Take the account of St. Andrews. At this juncture, Johnson barely mentions "one of the professors" who offers him lodging and does not name those involved in the "elegance of lettered hospitality" that greets him and Boswell (SJ 5). Instead, St. Andrews is treated as a site of "ruins of ancient magnificence" (SJ 5). Johnson takes the reader to the cathedral, the "fragment of the castle, in which the archbishop anciently resided," and the "university," which is examined as a "fabrick" (SJ 6, 7). Everywhere, Johnson walks among the "ruins of religious buildings," reading in them the history of Knox's reformation, the city's decline on losing its "archiepiscopal preeminence" (SJ 8, 6).

This focus on physical structures, on monuments and ruins, seems to match Johnson's attention elsewhere to distances and his willingness to correct other authors. Lauding Pennant's "delineations, which are doubtless exact" and claiming, quite emphatically, that "[m]ore nicety . . . is better," Johnson, the Enlightenment traveler and scientist, corrects the "fabulousness and credulity" of Boethius's estimation of the length of Loch Ness, for instance, with fact (SJ 149, 146, 30).[40] Sometimes Johnson simply reports distance: "[t]he length of Raasay is, by computation, fifteen miles, and the breadth two" (SJ 60). This is the Johnson that Richard Schwartz sees as "working in an extremely practical context, testing accounts and relating the evidence of the senses," "shar[ing]," Schwartz believes, "the scientific and antiquarian interests of many of his predecessors and contemporaries."[41]

All true. But the extraordinary particularity of Johnson's accounts of experience in his treatment of aesthetics exists in this domain too. Thus, we should pair Johnson's oft-quoted wisdom that "no man should travel unprovided with instruments for taking heights and distances" with passages like this one, early in the *Journey*, at Aberbrothik (SJ 146). Watch as Johnson returns to "use" and uses it to open up the antiquarian gaze:

> The monastery of Aberbrothick is of great renown in the history of Scotland. Its ruins afford ample testimony of its ancient magnificence: Its extent might, I suppose, easily be found by following the walls among the grass and weeds, and its height is known by some parts yet standing. The arch of one of the gates is entire, and of another only so far dilapidated as to diversify the appearance. A square apartment of great loftiness is yet standing: its use I could not conjecture, as its elevation was very disproportionate to its area. Two corner towers, particularly attracted our attention. Mr. Boswell, whose inquisitiveness is seconded by great activity, scrambled in at a high window, but found the stairs within broken, and could not reach the top. Of the other tower we were told that the inhabitants sometimes climbed it, but we did not immediately discern the entrance, and as the night was gathering upon us, thought proper to desist. Men skilled in architecture might do what we did not attempt: They might probably form an exact ground-plot of this venerable edifice. They may from some parts yet standing conjecture its general form, and perhaps by comparing it with other buildings of the same kind and the same age, attain an idea very near to truth. (SJ 11)

This is a long passage, dedicated to the kind of ruins that are bound to interest an antiquary. Johnson begins by gesturing toward the ruins as important in the "history of Scotland." King William the Lion had founded the abbey in 1178, and, like many buildings Johnson witnessed, it had been stripped in the Reformation of the sixteenth century. But Johnson gives us this context very briefly and does not linger here. His interest, rather, is in the building, its function, its "use." How is one to know from these ruins what the different parts of the structure were used for? We might expect Johnson to consult a source, either a book or a person on site. He does not. He is clear that "its extent might, I suppose, easily be found by following the walls among the grass and weeds, and its height is known by some parts yet standing." We might expect him to provide these measurement numbers; he does not. His focus, rather, is on how the ruins might be measured: "by following the walls among the grass and weeds."[42]

Indeed, although Johnson dangles the possibility of a factual result, the passage focuses instead on a variety of ways in which various spectators experience the structure.[43] We might say that the question about use seems to be answered not by a particular historical use but by the different ways in which the structure can be used and experienced. If measurement is the mode Johnson imagines for himself, Boswell presents another possibility. He possesses an "inquisitiveness seconded by great activity": he explores the physical structure, thrusting his body through a window. There are, too, the natives of Aberbrothik, whose relation to the structure is different yet: they "sometimes climbed the towers," the men are told, apparently able to perceive through repeated contact with the structure and shared communal knowledge an entrance that remains hidden to the travelers. Finally, Johnson describes for us a "skilled" perspective on the ruins, someone who might be able to draw them and reason from them, perhaps the closest to that Baconian traveler that Schwartz finds in Johnson himself.

Here, close to the beginning of the journey, Johnson calls the bluff of anyone who might point to travel as simply a record of "experience." What kind of direct experience would such an assessment have in mind, the passage provokes us to answer? Would it be the experience of the measuring antiquary, a more active and bodily experience that intrudes on its environment, an experience of daily use and custom, or an experience born of "skilled" sight and visual representation? In general British empiricism was not so invested in sorting out these differences. But Johnson notes them here and attributes them to various actors and communities; he describes them as discrete ways of knowing. A history of philosophy might consider such refined distinctions between forms of experience to wait until phenomenology centers them, but we find them clearly laid out for us in 1775.

This is not prescience but the contemplation of the craft knowledge at the heart of empiricism itself, and it moves beyond the "eye." It is as though Johnson takes the Baconian knowledge process he describes in the passage on aesthetics and chooses to focus not on its result—the "realities"—so much as on the ways of accessing them, on the process, then, as the main event. (We saw him do this with the pastoral and phantoms examples in the earlier passage.) Johnson shows us this complexity by breaking open the perspective of the antiquary to forms of experience that look more similar to those he contemplated in those passages devoted to perception and the imagination. As in those earlier passages, the "eye" is central to Johnson, though here his emphasis is on its limitations. In the scientific traveler's accounts of measurement, including Johnson's own, it can seem as though one practically sees distance or height or breadth. Johnson's concern is not just with the how of experiencing the structure but with the limitations of vision alone in accessing it. Boswell enters the structure, spurred by curiosity; the inhabitants enter the structure permitted by a special vision informed by use over time and communal understanding. Even the task of measurement is contemplated in terms of the kind of grass through which one would need to walk.

The limitations of the "eye" are everywhere in the *Journey*. Take this example, outside Inverness, of the first glimpses of the Highlands. Johnson begins, "Of the hills, which our journey offered to the view on either side, we did not take the height" (SJ 38). Here, as in the earlier passage, Johnson signals one way to record experience of the landscape but takes another tack. In similar fashion, he seems to set aesthetic response aside, keen to remark that he "did not see any that astonished us with their loftiness" (SJ 38). No sublime, then. He remarks that "Towards the summit of one [peak], there was a white spot, which I should have called a naked rock, but the guides, who had better eyes, and were acquainted with the phenomena of the country, declared it to be snow" (SJ 38). Here, Johnson seems to find the groove that suits his reporting best: not what he can see but the limitations of his vision and the skilled sight, at once "better" and also more informed, than his own. He launches into yet another account of hypothetical measurements—these of the heights of the mountains, "philosophically considered," which cannot be taken from the naked eye but must be "reckoned from the place where the rise begins to make a considerable angle with the plain" (SJ 38, 39). This method of measurement might, then, compensate for the limitations of the sight, which again proves inadequate to the task at hand.

Elements of this passage reappear in Johnson's account, now firmly in the Highlands, about caves off the coast of Mull. Here, Johnson does provide us with a numerical measurement: "more than a hundred and sixty yards, the eleventh part of mile," he says, is the distance the party has proceeded into the interior of the caves (SJ 146). But this proves not to be the half of it. We are told, for instance, that "though we went to see a cave, and knew that caves are dark, we forgot to carry tapers" (SJ 145–46). As a result, the measured distance is only what could be covered without light, rather than the measurement of the cave in a more thoroughgoing sense. Moreover, Johnson enlightens us, after giving us the number: "Our measures were not critically exact, having been made with a walking pole, such as is convenient to carry in these rocky countries, of which I guessed the length by standing against it. In this there could be no great errour, nor do I much doubt but the Highlander, whom we employed, reported the number right" (SJ 146). This seems very close to a parody of an account of "exactitude," that a critic like Schwartz is determined to show Johnson upholding.[44] But Johnson's other efforts at measurement give us context to understand it as something else: another example of his commitment to process, not to the measurement but to how it was taken and how it was recorded.

We can see here Johnson's direct engagement with the science of measurement of his historical moment. The midcentury had seen several important moves toward standardization. In 1742 the Royal Society approached the Royal Academy of Sciences in Paris to propose exchanging copies of the English standards of weight and measure for French ones. Among the items sent to Paris were "two brass rods, about 42 inches in length and ½ in. broad with squared ends." Behind such standardizing moves lies a messy set of previous standards, still stored and available for consultation. R. D. Connor describes how George Graham, a fellow of the Royal Society, was asked to compare no fewer than four discrete physical standard rods measuring the yard: "the Exchequer yard and ell of Elizabeth and the matrix into which they fitted, the Exchequer yard of Henry VII, the Clockmaker's yard, and the Tower yard." Keeping in mind, as Connor notes, that the Elizabethan rod's ends "were not parallel," these were Graham's discoveries:

1. The matrix of the yard exceeded the Elizabethan standard by 0.0102 inches.

2. The yard, (E), inscribed by Graham on the Royal Society's rod, exceeded the standard by 0.0075 inches.

3. The Henry VII yard was short of the standard by 0.0071 inches.

4. The Elizabethan ell exceeded 45 inches of the standard by 0.0494 inches.[45]

In the process of attempting to tidy up these relative measures, Graham "laid off the length of the Elizabethan Exchequer standard yard on the central longitudinal line of the three which had been engraved on the Royal Society's bar. . . . This bar came to be called the Royal Society bar No. 41 and is so inscribed today."[46]

We should take these midcentury efforts as background for Johnson's measurements in the cave. Just as he makes us question what it would mean to perceive a ruined abbey, Johnson opens up measurement itself, prying apart the relations underlying its apparent facticity. Instead of allowing us to accept the number as a given, Johnson asks his reader to work backward with him, from the number of the measurement to how it was made. At the moment when experts attempt to move toward international standards for measurement, Johnson pushes hard in the other direction, toward his own body. The body is manifested two times over in this account. For it is not only that Johnson has used a walking stick (a tool already gauged against the body) as an instrument to measure distance but that the stick itself has been established for measurement through its comparison to the body's height. Measurement in this passage is embodied, actively bringing together the measurement of the cave with the other work the stick does as a tool: moving the body forward, helping it to walk.

We can pull Johnson's account of measurement back into the Royal Society context that he features and which has been assumed to define him. We might think, for instance, of Joanna Picciotto's claims that Royal Society instrumentation serves as a way to purify the body, making a remote and objective form of observation possible. Johnson's stick as instrument allows us to see the ways in which he pushes back at this form of knowledge, recovering the body it so easily casts aside, and indeed recovering the instrument as part of the body, not in order to erase it, but to make it a habituated part of its workings. No longer restricted to sight, measurement is grounded on and assimilated to the body's own movements.

I want conclude this discussion with another aspect of the same measurement scene in Icolmkill, this one apparently more trivial. As I mentioned at the outset, there is another aspect of the day's experience that affects the number Johnson produces as measurement: the men forgot torches. This calls our attention to the problem there would be in assuming this as a measurement of the cave; rather, what is recorded here, in all of its embraced inexactitude, is measurement of this particular journey into the cave without those torches, a kind of computation of space based on the limitations of sight. Here, again, Johnson is keen to show us how much sight or measure contains, how much its context matters. This is the flip side of his curiosity about how the native Scots can "see" better than he and Boswell can. In both cases, the interest is in how "sight"—such an apparently automatic and easy sense—is actually cut through by tradition, by circumstance, and even by the time of day. If writers like Robert Hooke were keen to show the way in in which an instrument could add to sight, Johnson focuses on the many ways in which ordinary sight falls short. In Johnson's account of natural philosophy, experience is rooted in the mind and body of the observer and in the "horizon" (to use a much later term) of their experience.

The Useless Body

Having attended to the care with which Johnson integrates embodied experience into his account of Scotland, it is nothing short of jolting to read Boswell's description of the great man in the introduction to his published book.

> His person was large, robust, I may say approaching to the gigantick, and grown unwieldy from corpulency. His countenance was naturally of the craft of an ancient statue, but somewhat disfigured by the scars of that *evil*, which, it was formerly imagined, the *royal touch* could cure. He was now in his sixty-fourth year, and was become a little dull of hearing. His sight had always been somewhat weak; yet, so much does mind govern, and even supply the deficiency of organs, that his perceptions were uncommonly quick and accurate. His head, and sometimes also his body, shook with a kind of motion like the effect of a palsy: he appeared to be frequently disturbed by cramps, or convulsive contractions, of the nature of that distemper called *St. Vitus's* dance. (JB 18)

Whether or not this counts as the widely known Johnson that Boswell claims it is, the "character" sketch crassly refutes Johnson's account of embodied knowledge.[47] To that account, Boswell immediately poses these questions:

Which body does Johnson's account assume? Could this unusual body, this abnormal body, be the standard for anything? And how can an elaborate account of sensory experience be supported amidst the great man's "defects" (to use the eighteenth-century term privileged by Helen Deutsch and Felicity Nussbaum)?[48] Johnson's concerns with precision and measurement—his account of what he has been able to see and what not—now seem undermined by the knowledge that he cannot see or hear well at all. At least, that is the obvious threat that Boswell's opening poses, one that he seems to register in his excuse: "so much does the mind govern" that "perception" isn't an issue at all.

Although I began with the anecdote on Monboddo, in which Boswell showcases Johnson's philosophical remove from practice, Boswell's approach in the *Tour* is more complex. On the one hand, Johnson constantly evaluates the world, as Boswell puts it, "from his elevated state of philosophical dignity" (JB 14). He is, "the mighty sage," (or simply "Sage,") or the "grave philosopher" (15, 24, 261, 358). And, he is the "abstract scholar" whose "great mind," Boswell says, should not be "appl[ied] . . . to minute particulars" (JB 278, 145, 145). This is exactly the account of particularity we associate with Aristotle and the superiority of the philosopher's theory.

On the other hand, even as he holds up the sage, Boswell embodies Johnson to the fullest. Although Johnson's letters to Hester Thrale remark on his physical challenges, the *Journey* records almost nothing on this score.[49] With Boswell, however, Johnson's bodily limitations are always on view. As early as the trip to Monboddo's estate, Boswell records a worry about Johnson and the "tedious driving" of the day; at this point Johnson rather sassily reassures him, "Sir, I shall ride better than you. I was only afraid I should not find a horse able to carry me" (JB 84). But this, as Boswell documents it over the course of the trip, is a real worry. He describes for us how in the mountains "both Dr. Johnson and the horses were a good deal fatigued"; due to Johnson's being "a great weight," the guides agree that he will alternate horses (JB 144). At Rasay, Boswell tells us of a journey to the highest mountain on the island where the men do "a Highland dance," but Johnson stays behind, "unable to take so hardy a walk" (JB 168). It is frequently the case, as it is at Portree, that Boswell is in "a cordial humour," down drinking with the hosts, as Johnson "went early to bed" (JB 185). While the rest of the party walks, Johnson rides; while the rest of the party rides out a storm on the ship, Johnson rests, having already retired from sea sickness (JB 280).

Johnson's physical challenges may at first seem matters of fact only. After all, he was significantly older than Boswell, who could be understood as "shepherding an elderly man round a difficult and sometimes dangerous itinerary."[50] But such a view neglects the way Boswell moves seamlessly from this sort of mild account of "defects" to more impactful ones. I take the term *defects* from Helen Deutsch and Felicity Nussbaum's account as "both a cultural trope and a material condition that indelibly affected people's lives."[51] Notice that Boswell's "character" of Johnson alone meets all forms of the variety contained in this definition: "Defects including having one of the senses impaired—blindness, deafness, or the inability to speak—as well as physical anomalies such as being lame or possessing a hunchback."[52] Boswell makes of Johnson's body a veritable laundry list of such anomalies; it is the location of sensory impairment (seeing, hearing), as well as physical abnormalities (scarring, convulsions).

Lennard Davis points to Johnson in particular as a figure who evidences the new eighteenth-century appearance of "the disabled person in print as author and character," even as he exists in a liminal moment, straddling two different historical models of disability.[53] Historically we are on the way from an understanding of disability in terms of wonder to "an institutional, medicalized apparatus to house, segregate, isolate, or fix people with disabilities" (and its concurrent "clinical gaze").[54] Disability is moving toward a group identity, though it is not there yet, and biography would come to explain its subject through disability, rather than "downplaying" it, as Davis records Johnson's eighteenth-century biographers, Boswell and Thrale, doing.[55] In her focus on the cultural record of Johnson's unusual body during this liminal moment, Helen Deutsch finds a complex working out of broad social concerns about authorship and style through Johnson's body and its movements.[56]

One text that historians of disability consider of particular importance for the eighteenth-century move toward considering disability in terms of group identity is William Hay's *Deformity: An Essay* (1754). Hay, a member of Parliament with spinal deformities, is an unusual upper-class advocate for the humanity of the "deformed." One aspect of Hay's account that is of particular interest for Johnson's centers around the concept of "use." Hay observes that he is lucky indeed not to have been born in Sparta, where he would have been cast out as a "useless thing."[57] He says that his own caretakers, going hard in the other direction, have "out of Tenderness, tried every Art to correct the

Errors of Nature."[58] But this does not work, and Hay responds to its failure by switching metaphors. If his family's care suggests a tree trunk to be straightened, Hay seems more comfortable in explaining his own "defect" through a mechanical, functional metaphor, where solutions are not imagined as going back into nature but in compensating for it. In explaining what he calls the "certain Consequences of Bodily Deformity," he turns to use to explain his own "warped" and "disproportioned" "Human Frame."[59] Such a frame, he asserts, is "lessened in Strength and Activity; and rendered less fit for its Functions."[60] One must, then, render the body more functional, something easily imagined in these terms. He directs us, for example, to Scarron, who "had invented an Engine to take off his Hat," along with Hay's own desire to have something to tie his shoes and to pick up objects from the floor, like a lady's fan or glove.[61]

We see already the way that Johnson's projection of "use" back on bodily experience would have new stakes for a body that does not "function" well. Here, too, Johnson and Hay show us the extent to which, even as these broad cultural shifts are made visible through the body of Samuel Johnson, so too is a new kind of thinking about experience: specifically about the body's role in it. As disability historians and theorists, including Davis, have long indicated, "disability" is the name for a "socially constructed" relation; it is what happens when the environment makes an "impairment" matter, turning it into a "disability."[62] It is not surprising that Johnson and Boswell, two writers intently concerned with the relations of their own bodies to the places and people they observed, would begin to interrogate this construction. Look what happens to the walking stick, which operated in Johnson's text as part of the body's relation to the world, with the body's "defects" in mind. Near the end of the *Tour* Boswell stops to explain the object. He begins by noting that Johnson was "out of humour to-day" and this is due to his having "suffered a loss" (JB 318).

> The loss that I allude to was that of the large oak-stick, which, as I formerly mentioned, he had brought with him from London. It was of great use to him in our wild peregrination; for, ever since his last illness in 1766, he has had a weakness in his knees, and has not been able to walk easily. It had too the properties of a measure; for one nail was driven into it at the length of a foot; another at that of a yard. (JB 318)

Boswell, like Johnson, notes the way body and natural philosophy come together in the walking stick, but his account departs from Johnson's own entanglement of the stick and body as the source of his measurement. When we look at Boswell's description, which Johnson would have perused, the stick now looks as though it has been mystified by Johnson in his account of the relations behind its measurement: it turns out, per Boswell, that there was just a nail in the stick; this was a rather ordinary measuring stick, divided into measures of foot and yard. Moreover, the stick is now very clearly attached to the failure of Johnson's body, in particular, notes the biographer-in-training, to "his last illness in 1766," which produces "a weakness in his knees." This measuring stick, then, compensates for and tells the story of a body that is injured and unable to operate well on its own.

Boswell puts Johnson's "defects" on display. He also makes a connection that Deutsch and Nussbaum, following Rosemarie Garland-Thompson, do between these physical abnormalities and the monstrous. Garland-Thompson's account of early modernity emphasizes a turn away from the novel and toward Baconian science as a moment when "wonder becomes error."[63] In Boswell the "gigantick" of the character sketch is perpetually edging into the supernatural realm of the giant. The great man is described as a "giant among the luxuriant thistles and nettles," and again appears a "giant" when at play with a child (JB 55, 87). He is described as a "magnificent Triton," a "venerable *senachi*" (Scottish bard), an "*ancient Caledonian*" (JB 162, 324). He is also described as a whale, the planet Venus, and the constellation "Ursa Major" (JB 344). We can read in these metaphors the extent to which Johnson is strange and otherworldly: like all monsters, strange here but fitting in some other world, perhaps one beyond earth in the night sky, or in the zones of myth or history.[64]

We can also see in Boswell's treatment of Johnson's body a description of defect that brings with it a full awareness of the way cultural norms operate. Boswell is aware of the way Johnson's anti-Scots sentiment works, and he shows us how studying the great man is not exactly in opposition to studying society but merely another means toward that end. As he puts it, lightly but tellingly, in response to Johnson's jesting "at some deficiency in the Highlands," "the truth is, that if he had always worn a nightcap, as is the common practice, and found the highlanders did not wear one, he would have wondered at their barbarity" (JB 323). Barbarism, as Boswell points out here, is a construct, a way of marking a deviation from the norm. This kind of comment, the joking about Johnson's monstrous forms, suggests in Boswell a kind of reverse othering, pushing back at Johnson's constant diminishment of Scotland and its people and characterizing the knowledge that does so as monstrous. If, as Gordon Turnbull suggests, Boswell paints his relationship with Johnson as a kind of restaging of the 1707 Union, then he also shows us that there is a monstrous side to the otherness to which he bonds himself, a union that is ill-fitting, even a bit dangerous.[65]

Let's pause a moment and return to Boswell's accounts of Johnson as "philosopher" and "sage." Boswell gives us Johnson as a split figure: a theoretical mind that is associated with the philosophical and the abstract, along with a "gigantick" body that moves uneasily and with difficulty through the course of the journey. Midway through his physical "character" sketch of Johnson, Boswell assures us that the great mind compensates for the body. But if we think of Johnson's account, we realize that the damage is done. For what Boswell gains in his representations of Johnson's very materially present body is to disable the ease of the mind-body relationship that Johnson has labored to demonstrate. Here, in Boswell, the body is a burden, something that only the greatest mind can overcome. If it does so, its aim will be to leave that body behind, as its success counts as managing all experience-based perception on its own.

Monstrosity pushes this point much further. With the monstrous, Boswell calls up the potential for Johnson to work against the empirical knowledge project itself. We could think of this in terms of the category of the monster that Bacon tries to normalize, to make scientific and human, but that just keeps intruding into the sphere of this new empirical world, unresolved and thus an even bigger threat. This is the Enlightenment fate that Andrew Curran and Patrick Graille describe for the monster: "the anatomical corroboration of the breakdown of objective truth."[66] But, if we stick with Boswell, we don't really need Curran and Graille's elegant formulation. We might simply ask: What happens when the "sage," the "high priest of . . . reason," is understood as monstrous, as puncturing the very world that his embodied form of knowing makes possible? This is a criticism of Johnson and a critique of imperial knowledge. It is also a kind of ground clearing, a detachment of the traveler's experience—of experience, generally—from what is now contained within the unruly person of Samuel Johnson: the abnormal body, the philosophical distance, and the power and reason that subtend his knowledge world. When we recognize this, we can better see what Boswell offers us instead.

It is more—and different—embodiment. If Boswell leads us to see that in much of the journey Johnson's body is abstract, made invisible, one effect of highlighting the massive impediment that is Johnson's body is to bestow on Boswell's body a striking ease of movement. Indeed Johnson's formulation at the Aberbrothick abbey appears to take off from Boswell's own account of his knowledge-seeking activity. In Sky, Boswell's approach to the society is the equivalent of scrambling over walls and ruins:

> We danced to-night to the musick of the bagpipe, which made us beat the ground with prodigious force. I thought it better to endeavour to conciliate the kindness of the people of Sky, by joining heartily in their amusements, than to play the abstract scholar. I looked on this Tour to the Hebrides as a copartnership between Dr. Johnson and me. Each was to do all he could to promote its success; and I have some reason to flatter myself, that my gayer exertions were of service to us. Dr. Johnson's immense fund of knowledge and wit was a wonderful source of admiration and delight to them; but they had it only at times; and they required to have the intervals agreeably filled up, and even little elucidations of his learned text. (JB 278)

Note how Boswell consigns Johnson to the role of the "abstract scholar" and renders even his social interactions as "learned text." To Johnson's retreat, his distance, Boswell posits his very different way of knowing: one that is disposed to "join[] heartily," to be in the middle of the action. Boswell speaks of the relationship here as a "copartnership," with Johnson devoted to the role of the philosopher, Boswell acting, filling up space, and rendering the text understandable.

In Boswell, then, we see a formulation we would not expect from an eighteenth-century text: an account of a kind of participant observation, albeit one that involves two bodies rather than one. In this Boswell seems to theorize the unique kind of knowledge making the two can do together, claiming an epistemological sophistication that is usually thought to accompany the rise of anthropology as a discipline in the nineteenth century.[67] Over the course of the journey, however, Boswell shows us exactly what is necessary for this perfect epistemological partnership to succeed: the fabrication of a body that can disappear into the scene, becoming a part of it. In this, Johnson's highly materialized body serves as a foil for Boswell's own "useful" body.

Although Boswell's early formulations might lead us to suspect that Johnson occupies the sage or philosopher role and Boswell himself the practical one, this is not the way Boswell makes his argument. Instead, the two inhabit different versions of the practical and corporeal. Again and again, Johnson seems suited to the role of the abstract scholar, simply because his body is not up to the task of Boswellian exuberance. This is particularly marked in Boswell's multiple accounts of dancing. When the two are at Rasay (after the Highland dancing that takes place on the mountain), "a fiddler appeared, and a little ball began" (JB 166). Boswell relates of Johnson that he is able "to observe him sitting by, while we danced, sometimes in deep meditation—sometimes smiling complacently—sometimes looking upon Hooke's Roman History—and sometimes talking a little, amidst the noise of the ball" (166). Boswell runs through three different sorts of remove, with varying relations to the social setting, but in all of them, Johnson is *there*. He is not a remote, philosophical observer. We know the placement of his body, the books he picks up and puts down, the way his voice is covered by "noise."

As in Hogarth's turn to dance and "action," Boswell finds in dance an embodied sociality, a way to know the society of which one, however temporarily, is a part: "We performed, with much activity, a dance which, I suppose, the emigration from Sky has occasioned. They call it *America*. Each of the couples, after the common *involutions* and *evolutions*, successively whirls round in a circle, till all are in motion; and the dance seems intended to show how emigration catches, till a whole neighbourhood is set afloat" (JB 277). Boswell does not just come to know about emigration to America from the people who have watched their families and friends emigrate—though he and Johnson both obtain knowledge in this way. In addition, though, Boswell participates in this allegorical dance, becoming part of its "motion." In so doing Boswell's own identity—and certainly any distanced perspective on the scene—is lost, becoming part of the "all."

We can return to Gilbert Ryle to think further about why Boswell's accounts focus so minutely on dance. In explaining the difference between "knowing that" and "knowing how," Ryle puts his account of "performed" or embodied knowledge this way: "Intelligently to do something (whether internally or externally) is not to do two things, one 'in our heads' and the other perhaps in the outside world; it is to do one thing in a certain manner. It is somewhat like dancing gracefully, which differs from St. Vitus' dance, not by its incorporation of any extra motions (internal or external) but by the way in which the motions are executed."[68] We recall Boswell's own reference to St. Vitus's dance in his character of Johnson and account of the great man's "convulsive motions." Ryle helps us to see that Boswell marks off something quite different for himself than the "copartnership" he claims with Johnson. Rather, here, as established through dance, he offers his own distinct brand of embodied knowledge, one that ties intellect to action and that flourishes when mind and body work as one. This embodied participation, we should note, has none of the awkwardness of Banks's mourning ceremony, which turns the body over to the ceremony. Here, Boswell remains in control, reaching out with his authoritative, able body in order to access a social meaning that usually would be beyond him.

Given the way ideas are exchanged on (and after) the journey, we should be wary of treating a view on either side as a win or a conclusion. Let me demonstrate this by moving back to Johnson. If Boswell is interested in pressing on the register of experience that Johnson interrupts, Johnson has responded indirectly, by expanding the range of what constitutes sensory experience to begin with. We can find such an account in his description of Braidwood's Academy for the Deaf, very near the end of the *Journey*. To the modern reader this stop may seem peculiar, but, as Lennard Davis explains, there is nothing unusual about an eighteenth-century writer giving the deaf pride of place in an examination of humanity. In Davis's history of disability, "the deaf person became an icon for complex intersections of subject, class position, and the body."[69] This centrality turns on the Enlightenment fascination with language: "Deafness, after all, was about language, about the essential human quality of verbal communication. While Denis Diderot wrote on both the blind and the deaf, he saw blindness as posing a fundamental question about the nature of perception, whereas deafness was more fundamentally about the experience and function of language."[70] This was the age of the emergence of deaf academies in England, the first publications of books on deafness and the education of the deaf. Philosophical luminaries, including the historian of language, Monboddo, visited the Abbé de l'Épée's public displays of deaf students in Paris.[71] Central to Davis's claim about the purchase that deaf identity has on Enlightenment subjectivity is its focus on the textual, the discursive. As he puts it, given a written text, there is little difference between a hearing person and a deaf one in the reading or writing process; in this view, "Writing is in effect sign language, a language of mute signs." In this world, "the deaf person became the totemic representation of the new reading public."[72]

This account of deafness works well with the way that the Braidwood's Academy passage at the end of Johnson's *Journey* has been read.[73] There, Johnson, the great disparager of oral history and culture, lauds the deaf students relation to language, much as he had applauded the Scots' knowledge of English grammar gained by books:

> It will be readily supposed by those that consider this subject, that Mr. Braidwood's scholars spell accurately. Orthography is vitiated among such as first learn to speak, and then to write, by imperfect notions of the relation between letters and vocal utterance; but to those students every character is of equal importance; for letters are to them not symbols of names, but of things; when they write they do not represent a sound, but delineate a form. (SJ 163–64)

In language that harkens back to Sprat and Bacon, Johnson describes the deaf as having the best language-thing relationship one can imagine. Taking spoken language out of the picture gives language clarity and makes these students the ambassadors of the written language and history that Scotland, according to Johnson, historically has lacked.

But this is not all that Johnson says of the deaf students, and in an earlier passage he gives a suggestion of his reading on the topic: "I do not mean to mention the instruction of the deaf as new," Johnson begins. "Having been first practised upon the son of a Constable of Spain, it was afterwards cultivated with much emulation in England, by Wallis and Holder, and was lately professed by Mr. Baker, who once flattered me with hopes of seeing his method published" (SJ 163).[74] From this minihistory of publications on the topic, Johnson moves to his opening account of Braidwood's students:

> They not only speak, write, and understand what is written, but if he that speaks looks towards them, and modifies his organs by distinct and full utterance, they know so well what is spoken, that it is an expression

scarcely figurative to say, they hear with the eye. That any have attained to the power mentioned by Burnet, of feeling sounds, by laying a hand on the speaker's mouth, I know not; but I have seen so much, that I can believe more; a single word, or a short sentence, I think, may possibly be so distinguished. (SJ 163)

Before Johnson gets to writing, he moves through speech, but this speech is not limited to the senses we might expect. Instead, Johnson reports on those who hear with eyes and with their hands, compensations not of the mind but of the body. Even Johnson's own speech in the scene works this way. In order for the exchange to work, the speaker "modifies his organs" in speaking. He embodies the speech, making it clearer and more distinct in the way that he moves his mouth. There is a kind of parity, then, in the way some "organs" get modified, some switched out, in this exchange. And if we look at this paragraph, rather than the one on orthography, as representative of Johnson's views, a different account emerges: an account of sign language that is focused on the body, rather than on the reduction of signs to written language. Johnson highlights bodily supplements for speech and hearing, both.[75]

This scene takes on another meaning when we imagine Johnson performing and embodying speech with his own "defective" senses. He offers us a way of adding an additional valence to Davis's still-provocative claim that the Enlightenment was deaf. Here we might pull back to Johnson's insistence that senses are always compromised, in greater or lesser ways, and we can reread Johnson's account of the missing tapers in the cave as one instance of what Davis refers to as "universalizing marginality."[76] What meaning could it have, after all, to talk about defective senses, when sensory experience is so contingent, the environment and the body both constantly setting limitations or horizons on what can be known? In this sense, then, I think we could imagine Johnson speaking back to Boswell's characterization of sensory impairment and even to its account of a compensatory great mind. In Johnson's passage on Braidwood's, the senses compensate for the senses. In his account of the tapers, we are reminded that sensory impairment—as Johnson may have felt in his visit to Braidwood's—is a matter of degree and of circumstance. Sensory experience is hardly a given, always instead about how your body interacts with the environment.[77] Johnson could have come to this easily. After all the experience of disability is a profoundly situated one, attentive to the way in which environment presses on the body, to the ways that it enables or disables that body.

I have separated Johnson's account of the body compensating for the body from Boswell's account of the triumphant mind. But it is worth noticing that both of them play into the compensatory possibilities that view the disabled body as knowing better.[78] Both Katie Trumpener and Jason S. Farr show us this logic in the period's accounts of Duncan Campbell, a "deaf prophet" who relocated from the Highlands to London in the first decade of the eighteenth century. As Trumpener puts it, "the link between blindness and memory, physical handicap and sensory or mental compensation, appeared innate."[79] She describes how Campbell's "clairvoyant powers function both as a resistance against imperial occupation and as a compensation for the handicaps and disenfranchisements of history."[80] Farr describes Campbell's second sight as part of his "extrasensory capabilities."[81]

The recognition by both Trumpener and Farr that Campbell's limitations and his powers seem supernatural but are understandable in very natural terms (those of cultural criticism or disability theory) resembles eighteenth-century arguments for second sight, which also bring the supposedly superstitious practice in line with enlightened eighteenth-century philosophical considerations of experience. In his 1703 text, Martin Martin refers to second sight as an instance of the "supernatural."[82] Martin claims second sight is "a singular Faculty of Seeing," and he uses familiar empiricist terms to describe the "lively impression" it makes on the "Seer."[83] Johnson follows this, citing a "receptive faculty" and describing it in terms of an "impression" (SJ 107). In this part of the text Johnson has already dismissed as inaccurate perceptions of Scottish beliefs about agriculture as "superstitions," when, he says, they instead "regard only natural effects," comparing their attention to the moon to the kind of advice given in "English Almanacks" (SJ 107). His account of second sight, then, is meant to accomplish much the same thing, demonstrating that what might be assumed to be superstition is in fact understandable as a natural process. Indeed this "mode of seeing," Johnson wants us to understand, is "superadded to that which Nature generally bestows" (SJ 108). His most extraordinary terms for this come in his comparison of the second sight to cognition, something Martin seems also to have assumed but did not state so emphatically as this: "the second sight is only wonderful because it is rare, for, considered in itself, it involves no more difficulty than dreams, or perhaps than the regular exercise of the cogitative faculty" (SJ 109). To further his case, and to call our attention to the proper "natural" context for this observation, Johnson notes that "particular instances have been given, with such evidence, as neither Bacon nor Bayle has been able to resist" (SJ 109). He is not the first natural philosopher to think of this in terms of the work of the mind. When Johnson claims that this could be a "regular exercise of the cogitative faculty," he illuminates the logic that Boswell revealed for us: an unusual body could be the occasion for a strikingly different—and enhanced—way of experiencing the world.

It is humbling and instructive to try to account for Johnson's and Boswell's formulations of embodiment in terms that make political sense now. They do not line up well. Indeed, Johnson comes to think about his experience of the world through rhetoric and conceptions of "use" that began as an account of English superiority, a know-how that reinforced and help to make an imperial modernity. We should not balk at this or fail to acknowledge it. This is a messy history, one whose terms suggest both rigorous thinking that could and did seed possibility, even as it also implores us not to take for granted the political valence of a formal opening.

SIX

Fetish

EQUIANO'S ANTISLAVERY CRAFT

IN THE WORLD OF THE Enlightenment fetish, handicraft is not merely a description of process or product: it is an accusation. Bruno Latour, following William Pietz, describes the European concept of the African fetish as the result of a cross-cultural collision that produces belief as misunderstanding.[1] Its logic casts the "idol" or amulet worn by the African person not as evidence of a different belief system but as evidence of an incorrect way of formulating belief, one that is unable to comprehend the location of power. Latour voices the modern objection this way: "You can't say both that you've made your own fetishes and that they are true divinities; *you have to choose*: it's either one or the other."[2] Following Johnson's account of the poorly made objects in Scotland, we are prepared for craft to be weaponized, used as a diagnostic tool for ranking societies or civilizations. But the logic of the fetish uses making to go further yet. Onto the unwillingness to discriminate between the made and the given, the power located inside or beyond the object, Latour's moderns project inability in the form of ignorance, naïveté. The misunderstanding of the agency involved in craft opens up to a self-contradiction so deep as to dehumanize beyond a point of comparison to the European other.

My reading of *The Interesting Narrative of the Life of Olaudah Equiano, the African* (1789) centers Equiano's relation to the fetish and its potential for knowledge making alongside his commitment to a European account of craft. At first this will seem to pull in opposite directions: on the one hand, Equiano's commitment to dramatizing fetishes, African and European alike, breaks down the European superiority that so easily creates a belief grounded in a knowledge problem.[3] In his cross-cultural comparisons Equiano is interested in the supposed problem for the fetish: its trouble pinpointing the origin of its own power. On the other hand, Equiano is a craftsman himself, sailor and navigator, an expert in practical knowledge as European philosophy understood it: with a particular maker and consequent embodiment, the simplicity of which suggests the problem of the fetish in the first place.

In what follows, I read Equiano as a philosopher of practical knowledge, one who turns what the fetish has to offer on causation, power, and its relation to the body over and against the model of craft knowledge proposed by the Royal Society and disseminated, as we now have seen, through many different kinds of texts.[4] Equiano reminds us that the enslaved person's body is no easy location for embodied knowledge, given that it is the property of another. Furthermore, Equiano unpacks what we have thus far taken for granted about craft: its drive toward mastery. Craftsmanship is, as Richard Sennett puts it, "the skill of making things well."[5] This chapter examines the way Equiano opens up the concept of craft, turning back on its demand for mastery but employing its ideas for subjectivity in order to imagine the terms for a world after chattel slavery.

There is a significant history of scholarly disappointment at Equiano's lack of radical political proposals. As Geraldine Murphy puts it, "With postcolonial hindsight, we wince" at Equiano's glorification of commerce.[6] In this we would rightly contrast Equiano's approach with his friend Ottobah Cugoano's, whose *Thoughts and Sentiments on the Evil and Wicked Traffic of the Slavery and Commerce of the Human Species* (1787) ends with an enumerated list of social and political actions.[7] There is no doubt that Equiano proceeds otherwise, suggesting, not so much direct political actions his readers might take, as the ways in which they need to think in order for political change to be truly realized. We find ourselves in a similar position now. Decades after equal rights under the law, philosophers and literary critics have turned to approach aspects of Black subjectivity that seem unresolved through this history, indicating Equiano's prescience: a postslavery society must include a shift in how we all think about subjectivity and collectivity. Although Equiano stages himself as a bourgeois individual, the "self-made man" of the subtitle of Vincent Carretta's biography, the subjectivity he offers to us is profoundly divergent from this implicitly white Enlightenment model:

it is material, embodied, and shared.[8] He allows us to understand the new possibilities—and necessities—for a Black Enlightenment subject through the discourse surrounding craft.[9]

I argue, then, that *The Interesting Narrative* offers us a roadmap for how we should think about people and experience in the society that could exist after slavery. For Equiano, and for millions of other descendants of enslaved persons, epistemology could be a matter of life and death. Equiano is well aware that slavery rests comfortably on discourses like natural philosophy and natural history, which have a stake in giving us the world (natural, social) as it is. But he is also proximate to accounts of the investigation of nature as a matter of "making," not just of discovery. Like other authors in this book, then, Equiano elevates that craft aspect of eighteenth-century experience and pushes it to the fore. He sees the need for us to think about oppression, not only in terms of actions, but in terms of how we experience the world and the people around us.

Knowing Through the Fetish

Writers on Equiano have noticed that the terms for difference in *The Interesting Narrative* are not primarily those of skin color. Equiano, of course, shows us how Black skin can be a liability for the freed slave; and he shows us moments—with family members and strangers alike—where the Black body or face serves as a point of connection. On the whole, though, Equiano's antislavery arguments instead appear to engage the differences between Black Africans and White Europeans along the lines of the eighteenth-century debates over the status of African civilization. Roxann Wheeler identifies the central terms for Equiano's depiction of Africa as "historical and economic forces."[10] George Boulukos shows us how Equiano closely follows Anthony Benezet's influential antislavery account of Guinea, according to which African societies were understood not in "terms of race or nation, but in an older tradition, through their religious identity and state of civilization."[11] As his own copious references to Benezet suggest, Equiano is following the terms of an already vigorous antislavery debate, in which the humanity of Black Africans could turn on the newest accounts of what makes up society in the first place. Along these lines we find Equiano depicting Igbo society in the terms of Adam Smith's four-stages theory;[12] and we find writers on both sides of the slave debate quoting Montesquieu's *L'esprit des lois* (1748).[13]

Even as Equiano employs Scottish Enlightenment ideas of society as ways of conceiving of human difference, however, he trains us in ways to think difference that exceed, and even turn back on, the four-stage progressive agenda. To make his argument Equiano turns to the European conception of the fetish. Other critics have noted the presence of the fetish concept in the young Equiano's famous encounters with moving portraits and talking books in his early life on the slave plantation. But scholars have not connected that moment to the concept's presence in Equiano's early descriptions of Africa, which prefigures and complicates that later scene. As Equiano himself points out, the African passages are highly indebted to other antislavery writers' accounts of the continent; thus, it is not surprising that we might find articulations of the fetish running through such accounts, themselves hybrid creatures that borrow extensively from earlier travel narratives. In highlighting the ramifications of Equiano's deployment of "fetish" to describe African-ness, I draw on William Pietz's location of this concept with Enlightenment travel writing and his account of its great reach within the period's intellectual culture. Formulated on the Gold Coast of Africa, in "an ongoing crosscultural situation outside Europe" between the fifteenth and seventeenth centuries, the concept of "the African fetish worshipper was a paradigmatic example of what was *not* enlightenment."[14] By the seventeenth century, as Pietz describes it, the pidgin "fetisso" referred to everything from "African religious objects" to the "general socio-religious system of African peoples."[15] There are two aspects of Pietz's "problem-idea" of the fetish that will be especially useful for thinking about Equiano: First, its "irreducible materiality," its status as a "material embodiment," as opposed to the idol's "iconic" relation to the immaterial and the consequences of this for a "personhood . . . conceived as inseparable from their bodies."[16] This is literalized in the "amulets . . . worn by common people for certain magical effects" that often are the direct referents of "fetisso," themselves strangely material and strangely proximate to the body.[17] Second, its "mystery of value," and thus its calling out to the "nonuniversality and constructedness of social value" more generally.[18] The fetish, after all, is the thing whose value cannot be understood in terms of the "functional objects" of the mercantile code; it is, rather, troubled in its "transcoding" and thus, Pietz observes, revealing of "an anthropological problem of cross-cultural judgement determined by the question of the social value of material objects."[19]

The fetish as described by William Bosman in *A New and Accurate Description of the Coast of Guinea* (Dutch 1702; English 1705) traveled extensively through Enlightenment philosophy. Pietz describes the work by the onetime chief merchant of the Dutch West India Company as "the authoritative account of black Africa for eighteenth-century Europe."[20] It was plagiarized widely in other prominent descriptions of Africa of the period, such as Jean-Baptiste Labat's famous 1730 account of Guinea, and it was the leading account of the region in Thomas Astley's famous English voyage collection of 1743–47.[21] Moreover, as Pietz observes, Bosman's book made its way into the libraries of Isaac Newton, John Locke, and Edward Gibbon; it was argued over by Pierre Bayle and referred to by Adam Smith in his *Lectures on Jurisprudence*.[22] Bosman's text served as the foundation for theories of natural religion, present in David Hume and borrowed by Charles de Brosses, that aimed to explain polytheistic belief.[23] There is little surprise, then, that it might also find its way into Equiano's *Interesting Narrative*.

Equiano's depictions of African religion—and indeed of European objects in later passages—offer us accounts of belief filtered through the concept of the fetish. Equiano is especially concerned with the ideas of "making" that Latour finds so important to the idea of the fetish, the sense, in short that the *object* of belief is, to the European observer, troublingly constructed rather than given. Equiano is interested, too, in an aspect of the fetish that is deeply tied to this but is more prominent in Pietz's early account: the material object-body relation that leads to a strong conception of the "embodied" African.[24] Equiano sees very clearly the overlap between these disparaging views of African belief and society and the related—and very much lauded—views of English natural philosophy. As we will see, he frequently uses one set of terms to work through the other.

We might begin by noticing how Equiano takes up the logic present in Bosman's account and turns it toward surprising ends. Pietz observes that one of the chief aims of the discourse of the fetish is to insist on the bifurcation of society into manipulative priests and naïve believers.[25] In Bosman's account all African religion is manipulation, the product of false priests and (as he call them in Letter X), "Miracle-Mongers."[26] This, indeed, is the part of Bosman that appealed to a writer like Bayle, who was keen to expand on the relation between these false priests and those of Catholicism.[27]

We see Equiano approach the same articulation in his first chapter's account of his Igbo homeland, what we now know as southeastern Nigeria. Equiano describes Igbo "magicians" who "were also our doctors or physicians" (*IN* 27). While Equiano acknowledges the success of these persons in curing physical ailments ("healing wounds" and "expelling poisons"), he moves on to what he prepares his European readers to understand as more dubious claims: "some extraordinary method of discovering jealousy, theft, and poisoning" (*IN* 27). In this, Equiano observes, the "magicians" probably derived their "success . . . from their unbounded influence over the credulity and superstition of the people" (*IN* 27). At this point, Equiano offers "an instance or two" for the reader's consideration:

> A virgin had been poisoned, but it was not known by whom; the doctors ordered the corpse to be taken up by some persons, and carried to the grave. As soon as the bearers had it raised on their shoulders, they seemed seized with some sudden impulse, and ran to and fro unable to stop themselves. At last, after having passed through a number of thorns and prickly bushes unhurt, the corpse fell from them close to a house, and defaced it in the fall; and the owner being taken up, he immediately confessed the poisoning. (*IN* 27)

The credulity of those who believe in this "instance" is marked for us up front, of course, but the anecdote itself goes uninterpreted. If there is an abuse of power here, what was its mechanism? Which piece of the story should we doubt? Equiano offers a footnote, and (especially if we have read Bayle) we might expect it to be the locus of demystification. But what is offered there turns out to be something quite different: a second instance, this one at Montserrat "in the West Indies in the year 1763" (*IN* 27, n2). This event is grounded in Equiano's own experience, in fact something he has heard secondhand: "I then belonged to the Charming Sally," he tells us, and relates the experience as that of the "chief mate, Mr. Mansfield" (*IN* 27, n2). Mansfield and the other sailors begin in the position of skepticism that Equiano has set up for the reader in the main body of the text: "Though they had often heard of the circumstance of the running in such cases, and had even seen it, they imagined it to be a trick of the corpse-bearers" (*IN* 27, n2). Here, then, perhaps, the abuse of power we were encouraged to attribute to the "magicians" has been slightly relocated, with the coffin bearers themselves obviously behind the ruse. But the skeptical sailors are drawn into the ceremony and find themselves acting the very part of the coffin bearers. They raise the coffin to their own shoulders, only to see it fall against a hut, whose owner confesses to the poisoning. "The credit which is due to it I leave with the reader," Equiano remarks, signaling a hands-off approach that covers over his collapse of the positions—outsider/insider; skeptic/believer—that held up Bosman's account of superstition (*IN* 27–28, n2).

This is not the only work with the fetish and with superstition that Equiano does in the "ethnographic" sections of the work. In Letter X Bosman refers to "Oath-Draught[s]," the material communications with the spirits featured in the African polytheistic culture on which he reports.[28] Equiano calls them "libations," but their function is the same, as in his account of the "manners and customs" of the Igbo people (*IN* 21, 17). Listing them under the "manner of living," Equiano describes these offerings to the gods alongside the "cleanliness" he attributes to the everyday manners of the people: "After washing, libation is *made,* by pouring out a small portion of the drink, in a certain place, for the spirits of departed relations, which the natives suppose to preside over their conduct, and guard them from evil" (*IN* 21; my emphasis). I stress the "made" of the above description because we might think of the offering of libations in terms of belief, but Equiano's paragraph steers us additionally to think of them as common practices. This paragraph begins with "cookery," with the "bullocks, goats, and poultry" that the people consume and, further, with how they spice their food (*IN* 20). Equiano thus seems to offer us "libations" as such an everyday custom, its "small portion" and "small quantity" giving a recipe of a kind for the worship of spirits (*IN* 21).

It is hard to imagine a more demystifying context for "libations" than the placement of them alongside the salting of food and the washing of hands. This kind of move, of course, is what has made Equiano's account seem to many readers to be "ethnographic," comfortable in its treatment of religion as just another sort of custom. But this does not quite account for the way that Equiano brings the reader into Igbo religion. When "libations" come up again, just a few pages later, they are part of an explicit discussion of "religion." Equiano begins by reporting that the Igbo have "one Creator" but not "the doctrine of eternity" (*IN* 40). Then, he turns back to the matter of libations and spirits and connects it to the practice of offering "oblations of the blood of beasts . . . at their graves," at which point he reports his own experience of the latter, when he accompanied his mother to "her mother's tomb" (*IN* 25). As a boy, Equiano "sometimes attended her," but he does not participate in the "oblations": "I have been often extremely terrified on these occasions. The loneliness of the place, the darkness of the night, and the ceremony of libation, naturally awful and gloomy, were heightened by my mother's lamentations; and these, concurring with the doleful cries of birds, by which these places were frequented, gave an inexpressible terror to the scene" (*IN* 25). If we think back to Bosman's "Oath-Draughts," we notice that Equiano dramatizes an outside to superstitious practice that is not that of the skeptic. As in the example of the coffin carriers, Equiano insists to the reader that the fetish's conception of the naïve believer and the skeptical outsider may be insufficient to understand how belief works. In the case of the "ceremony of libation," the boy is outside the ceremony but also in tune with the "naturally awful and gloomy . . . lamentations." The young boy is in tune with his African mother and, simultaneously, with the terms of eighteenth-century aesthetic philosophy, as Equiano's language of "gloom" and "awe" might well have been taken from Edmund Burke's *Enquiry* or a Gothic novel. Equiano, we might say, takes advantage of that, showing us how well the position of the boy (outside the ceremony) and the mother (inside the ceremony) fit together.[29] A common account of Equiano's autobiographical narrative is that its "progressive enlightenment" replaces his earlier superstition.[30] We might expect it in this passage, especially when we consider aesthetic philosophy as a kind of secularized religion.[31] But this is not what Equiano gives us, instead using the structure of autobiography to juxtapose two positions, showing them side by side.

We can see this, too, in Equiano's explicit treatment of the fetish. There are no African fetish objects worn around the neck in *The Interesting Narrative*. When there is an obvious material fetish, it comes in the form of European commodities. After Equiano is taken to Virginia, to the plantation of his first master, Mr. Campbell, he is alienated from his fellow Africans. "I now totally lost the small remains of comfort I had enjoyed in conversing with my countrymen," Equiano begins the chapter (*IN* 45). Finally at work as the only African on the plantation, accustomed to "weeding grass" and "gathering stones," Equiano is called to the master's house to fan him as he sleeps:

> While he was fast asleep I indulged myself a great deal in looking about the room, which to me appeared very fine and curious. The first object

that engaged my attention was a watch which hung on the chimney, and was going. I was quite surprised at the noise it made, and was afraid it would tell the gentleman any thing I might do amiss: and when I immediately after observed a picture hanging in the room, which appeared constantly to look at me, I was still more affrighted, having never seen such things as these before. At one time I thought it was something relative to magic; and not seeing it move, I thought it might be some way the whites had to keep their great men when they died, and offer them libations as we used to do our friendly spirits. In this state of anxiety I remained till my master awoke, when I was dismissed out of the room. (*IN* 46)

The boy thinks that the "watch" talks and the portrait's eyes move, and is able to make a rather startling analogy: "I thought it might be some way the whites had to keep their great men when they died, and offer them libations as we used to do our friendly spirits." Henry Louis Gates, Jr. glosses the sentences this way: "Under the guise of the representation of his naïve self, he is naming or reading Western culture closely, underlining relationships between subjects and objects that are implicit in commodity cultures."[32] Lynn Festa pushes this point further, drawing it out and making it central to the way the passage employs the fetish: in this way, she explains, Equiano reveals "European misrecognition of their own proper fetishism."[33] For her, Equiano thus marshals the object as a form of critique: "Equiano's improper animation of the clock and the picture (mistaking objects for subjects) masks the improper reification performed by the Europeans who mistake subjects for objects in the form of slaves."[34]

Festa's account makes a lot of sense: in an antislavery argument, Equiano turns the objects back on the idea of slavery itself. And yet the passage moves another way, one that suggests less a turn toward analogy as critique than toward a kind of equivalence. Again, Equiano works with Bosman's account of the African fetish. Let's look more closely at what Bosman's letter contains on this score. If Letter X makes libations rather central, it also features the African fetish object, which is called an "idol" by Bosman's English translator.[35] Pietz describes for us the fetish's "irreducible materiality" and describes as one of the themes of fetish discourse the object's relation to "individuals whose personhood is conceived as inseparable from their bodies."[36] The fetish, he says, "was typically some fabricated object to be worn about the body," though Bosman's tenth letter depicts another kind of bodily relation.[37] There he informs the reader of a "Wooden Pipe filled with Earth, Oil, Blood, the Bones of dead Men and Beasts, Feathers, Hair."[38] Bosman follows the actions of the "Feticheer" as he takes some of "the mentioned Ingredients out of the Pipe; with which he touches the Swearer's Head, Arms, Belly and Legs," and, following this, places into the pipe "a bit of the Nail of one Finger in each Hand, of one Toe of each Foot, and some of the Hair of his Head."[39] Of the body, on the body: even if separate, not seen like an idol might be but touched—or touching. It is worth remarking that there are not any such pipes or "idols" in *The Interesting Narrative*, though surely Equiano would have been familiar with them not only from Africa but from his extensive time on both Caribbean and American slave plantations. Instead, he gives us the supposedly quintessentially African as the European.

Equiano recalls this very bodiliness and crafted materiality when he describes the objects in the scene on the plantation. He encourages us to think about the fetishes in relation to the body of the master, when he makes them all, man and objects, quasi-conscious things: the master sleeping, the clock and portrait "telling" and "looking." Second, in using the term *watch*, which the *Oxford English Dictionary* confirms was most often used to describe a time piece kept on one's person in this part of the century, Equiano makes us imagine the object "hanging" not on the wall but on the master's body—in a manner we associate with the African religious object.[40] Finally, in suggesting that we think about portraits as preserved "great men," Equiano doesn't just make portraits and spirits equivalent; he draws attention to the even more material—as well as deliberate, intentional—construction of the European painting. It is made, a crafted idol. At the same time, the comparison causes us to wonder what, exactly, would constitute the "libations," were we to see through the analogy. Could it be that looking is its own form of communication and that the body is more related to the act of perception than we might have guessed? Hogarth thought so and so too did Equiano, as we will see in a subsequent passage. No ordinary objects, clock and portrait suggest to us, not only that objects as well as subjects are sources of power, but the extent to which this power is enabled by a form of belief that is intimately associated both with society (they do that; we do this) and with the body.

Rather than impropriety and exposure, as Festa suggests are its critical terms, the passage points us toward a less obvious conclusion: one that stresses the connections between African and European forms of thought. Let me stress that this is a fairly unsettling thing to find in an antislavery text written by a formerly enslaved person. Put simply, why at a juncture that should not brook a hint of moral ambiguity would you introduce something that looks a lot like the later conception of "cultural relativism"?[41] It is possible, however, that this is the wrong question entirely. Critics have long assumed that Equiano's aim in the text is mastery: mastery of the self through literacy or through economics. As Houston Baker put it in a critical early reading of the text, for the "propertied self . . . economics *must be mastered* before liberation can be achieved."[42] Equiano buys himself, and he does so with wealth he accumulates through trade. At the same time, though, he shows us, through the logic of the fetish, that subjugation and mastery are not the only options; indeed, he demonstrates the extent to which the material object cannot be properly understood in relation to the self as mere property. This, after all, is what makes the fetish so stubbornly resistant to European thought: it is not reducible to commodity status. For him, *fetisso* brings together "semantically the essential problems not only of cross-cultural perceptions of social order, but of the transvaluation into commodities of objects that were social values."[43] Festa assumes that "subject" and "object" map onto "master" and "slave."[44] But what if they do not, as Equiano suggests in turning the European commodity-objects back into an African sort of fetish? What Equiano ultimately seems to be looking for here is a way around a simple reversal of that power dynamic.

Where we would expect Equiano to close down—critique the European ideas of the object—he instead opens out. As with the Gothic moment of African ceremony, the emphasis does seem ironic but not satirical, an exposure of underpinnings but not a barbed attack on one way of knowing or believing or another. Equiano offers us something that is less like the philosophes using the fetish to expose and ridicule Catholic practice, and more like Michael Taussig's anthropological juxtapositions, when he, for instance, sets Walter Benjamin alongside the San Blas Cuna shaman, mystifiers both, or when he urges us to think of mimesis as a way that "the reproduction of life merges with the recapture of the soul."[45] Or like Latour's analysis, when he speaks the African position in the fetishism encounter—but what about the amulets you wear of your Virgin?—and then goes on to offer from it a correction to the blindness of the "Moderns'" scientific ways of knowing.[46]

Equiano both inspires us to establish a comparative mindset—"libations" are like "portraits"—in a way extremely close to what later anthropology would suggest, and forces us to see that a full analogy changes the way we might think about the European side of the comparison. Put differently, Equiano wants us to see an "African" aspect of European fetishism that such fetishism—indeed the very concept of fetishism—cannot acknowledge for or about itself.

I am suggesting, then, that the critics who have called Equiano an "ethnographer" have been on the right track but have understated the case in such a way as to fail to grasp the deep enquiry in *The Interesting Narrative* into questions about difference.[47] Indeed, Equiano's engagement with the fetish reveals his text to be startlingly self-conscious of its own movements, often undercutting the assumptions on which prominent theories rest. We can no longer, for example, take Equiano's engagement with Smith's four-stages theory—his participation in debates over how African society measures up to European society—at face value. Despite his employment of a recognizable Enlightenment vocabulary, *The Interesting Narrative* is not ultimately "progressivist in mood and ideology," motored by the "civilizing processes" of "Christianity and commerce."[48] Equiano's employment of the fetish undercuts both the teleology of this progressive program and its focus on "property" in particular as a means of gauging development. We can see this if we go no further than Gates's evaluation of Equiano as able to "read these objects in both ways."[49] It is more striking still if we take stock of the consistent way in which Equiano levels European and African perceptions of the world by remystifying the object and, indeed, showing the demystifying "ethnographic" or critical position itself to be part of this remystification.

This last statement requires more explanation. Equiano makes his argument by adopting European ways of knowing and leading the reader to question her relation to the superstitious ways of knowing associated with the African. Sometimes readers have missed this due to their assumptions about the structure of the autobiographical mode Equiano uses. Gates—and many readers after him—treat the naïve and analytical positions Equiano voices as comprehensible within the structure of autobiography, as the production of present and past self.[50] Thus, Gates says that Equiano "can now read these objects in both ways, as he once did in the Middle Passage but also as he does today."[51] But the passage itself is stranger, and its verb tenses do not reflect this before-and-after formulation.[52] Rather, it stages both perspectives in the past and thus asks us to grapple with their simultaneity. The boy is superstitious and analytical at the same time, a position that makes little sense for autobiography but makes tremendous sense alongside the kinds of comparisons that Equiano offers us in these passages, his leveling-out of ways of knowing, his gentle probing of the superstition within Enlightened sight.[53]

We can see the irony with which Equiano approaches questions of difference. As the passage on the plantation should suggest, the placement of his rigorous questioning about knowledge is also no accident. Equiano places a similar interrogation of positions onboard the slave ship, giving us epistemological argument even as he describes scenes of abjection and dehumanization. Indeed, what might at first appear discontinuous should be understood instead in terms of Equiano's powerful practice of argumentation. Simon Gikandi describes the "gist" of the slave ship passage as "the violent transformation of the self from a being into a commodity."[54] At this juncture Gikandi invokes Orlando Patterson's term *social death* to describe this moment, and he returns to Hortense Spillers's account of the Middle Passage as producing persons "unhinged from communities of kinship."[55] For Gikandi, then, Equiano here dramatizes the extent to which he is stranded in "a state of non-culture."[56] This is a powerful account of the Middle Passage, one that Equiano seems both to endorse and, in typical fashion, also to undermine. For amid this scene in which great violence subjects enslaved persons to a debasement so severe that others control their eating and thus their survival, Equiano offers scenes of connection to other Africans. A "little time after," Equiano remarks, "I found some of my own nation" (*IN* 40).

At a moment when ties are inevitably ruptured, at what we might think would be the beginnings of what Orlando Patterson calls "natal alienation," Equiano reports that he has "found" kinship.[57] This, he says, "in a small degree gave ease to [his] mind" (*IN* 40). He then asks his "countrymen" to report on the situation at hand: "I inquired of these what was to be done with us" (*IN* 40). Following this, however, his line of questioning is rather less predictable: "I asked them if these people had no country, but lived in this hollow place (the ship): . . . 'Then,' said I, 'how comes it in all our country we never heard of them?' . . . I then asked where were their women?" (*IN* 40). And, finally, Equiano asks, "how the vessel could go?" (*IN* 40). It is these last two moves that interest me: the boy's inclination to ask about the social underpinnings of slaver society, and then the way he follows up that line of questioning with a question that will reliably produce evidence of superstition. How does the vessel go? The "countrymen" answer, "magic" (*IN* 40).

If we expect Equiano to shelter himself in the small comfort his countrymen can provide him, he instead turns outward, toward questions that attempt to understand the slavers as social beings. April Langley has described the way in which Equiano's journey toward the coast is "comparative" in the way it takes in a variety of African societies.[58] That same interest would seem to continue here. In asking about the slavers, of course, Equiano performs a clear reversal of the puzzled European responses to other societies familiar to us from the age's travel writing. The irony of putting such "enlightened" queries about society in the mouth of an enslaved boy is plain. Again, though, what one feels in reading this passage, more than anything else, is a leveling that is also a remystification. Even as he describes the reduction of his own subjectivity, the process of being made into a commodity, Equiano claims a kind of equal footing for his own perspective in the scene. But it is an equal footing derived from a line of questions that moves unproblematically from social theory to fetishistic magic, both perhaps (we are led to think) rather magical, riddled by troubles with causation.

We find a related move just a few paragraphs later, in an aside that seems so out of place it is rarely discussed in readings of this pivotal passage. Having just described the suffocating and deadly conditions for enslaved people in the hold—the crowding and "the stench of the necessary tubs, carried off many"—Equiano turns elsewhere (*IN* 59). "During our passage I first saw flying fishes," he reports in the next sentence, with an abruptness worthy of a logbook (or Banks's journal) (*IN* 59). Here, then, Equiano performs what is now a familiar move: just on the heels of the superstitious "wonder" the enslaved people express at the ship, he shows an early scientific source of wonder, a curiosity of the natural world. Again, Equiano takes advantage of the easy slip between natural philosophy and superstition.

Equiano further works out a relation between scientific pursuits and wonder when he relates his experience with a borrowed quadrant:

> I also now first saw the use of a quadrant; I had often with astonishment seen the mariners make observations with it, and I could not think what it meant. They at last took notice of my surprise; and one of them, willing to increase it, as well as to gratify my curiosity, made me one day look through it. The clouds appeared to me to be land, which disappeared as

> they passed along. This heightened my wonder; and I was now more persuaded than ever that I was in another world, and that every thing about me was magic. (*IN* 42)

Let me gloss the amazing logic of this scene, on the face of it a conventional encounter between "savage" and technology. We may return here to the African religious object's "irreducible materiality," to that idea of subjectivity that puts object and body very close together. To the argument about fetishistic superstition—that Africans experience the world differently because they make knowledge mistakes about it—Equiano seems to answer that the enslaved boy does see differently. What is this different and disorienting sight? When the boy sees the sailor's "make observations" with the quadrant, he sees them perform a kind of observation that would ultimately allow them to calculate the geographical position of the ship. The sailors would have been looking through and lining up two tiny rectangular plates with pinhole sights and focusing on the North Star; a pendulum-like string with a weight hung from the quadrant, and it would register the angle or elevation between the horizon and the star. The view through the pinholes, then, in the ordinary use of the quadrant was positional, a lining up, a way of making sure the device itself was in the proper position to register angle and therefore to allow the sailor to calculate latitude.

But when the boy looks through the pinhole sights at the clouds in front of him, he is not lining up and calculating; he is seeing. And what he sees is disorienting and wonder-producing: "The clouds appeared to me to be land, which disappeared as they passed along." Equiano concludes the description by returning to the idea that "every thing about me was magic." Here, we should think again of Taussig's account of mimesis and its creations of magic, available through a quadrant and a religious object, both. The young boy's sight is not a hallucination; it is, rather, a "misuse" of the scientific instrument that turns it into a different kind of viewer. The pinhole viewer that Equiano has effectively created makes for—we might even say, crafts—a grainy, material-seeming image.

Equiano is on his way to a slave market. He is on the deck of a ship in which enslaved people are dying. That might encourage us to see the quadrant moment as playful but somewhat out of place, trivial. But we should reflect on how truly radical Equiano is in what he permits to sight. Take Pietz's introduction of Bosman's early account of the fetish. When Pietz describes how early "voyages to non-monotheist lands raised the problem of the material object in a new way," he juxtaposes these suddenly rich theoretical objects, possessive of social value, to the "very production of a geography of navigation routes," which "charted measurements of distances and depths," a kind of knowing committed to "a world-picture of functional objects stripped of cultural meaning and social value."[59] I don't think this necessarily captures what navigation was or is all about, but I think that Equiano draws much the same contrast that Pietz means to here. When the boy looks through the view hole he does not "see" abstract measurement but a grainy orb, an image but so material an image that one can nearly reach out and touch it. The young Equiano uses a quadrant to create a fetish. Indeed, he embodies the observer and turns the idea of "making observations" back in the direction of craft.[60]

Equiano's example of the flying fish offers its own account of subjectivity and, similarly, by taking up the example of the scientific observer. Following the account of the "magic" of the ship's movements, we see Equiano offer the natural historian's wonder at this natural curiosity or marvel. But Equiano's aim here is not just to suggest that the scientist might be more superstitious or the superstitious more scientific than we thought. For as he makes this point, Equiano also draws our attention to the observing subject through the abrupt shift in his writing. How does one move from the death below ship to the fish jumping onto it? We might look at this as a kind of generic shift in a slave narrative riddled with other generic moments, here to natural history writing. Although this isn't entirely satisfactory—*The Interesting Narrative* is a generic amalgam that makes such parsing hard to justify—it does point us toward one aspect of the text that is especially jarring. Equiano has shifted the very constitution of the observer. We might say that our narrator turns cold: from someone who could faint at the expressions of the faces of enslaved people onboard the ship, he has become the relatively emotionless observer of nature. Even this distinction, however, seems barely to scratch the surface, for just a few sentences earlier, Equiano had informed us that enslaved people do care very much about fish and had in fact used the animal as an example of the "cruelty of the whites":

> One day they had taken a number of fishes; and when they had killed and satisfied themselves with as many as they thought fit, to our astonishment

> who were on the deck, rather than give any of them to us to eat as we expected, they tossed the remaining fish into the sea again, although we begged and prayed for some as well as we could, but in vain; and some of my countrymen, being pressed by hunger, took an opportunity, when they thought no one saw them, of trying to get a little privately; but they were discovered, and the attempt procured them some very severe floggings. (*IN* 41–42)

Following this description, Equiano relates the suicides of several of the enslaved, who jump over the nettings to their deaths. The flying and falling of the natural history "flying fishes," when that brief description appears, is disturbingly resonant with these earlier flying bodies, some of which are dragged back to the ship and severely punished. If our first response as a reader is some discomfort at the shift from the scene of slavery to the remove of the fish, then, Equiano fills out for us why this would be. And in the process he points us to what we might otherwise have normalized in the "objectivity" of scientific observation. Now this remove appears so unlikely—so painfully unlikely—as to stop us in our tracks. It seems unlikely, that is, because Equiano has forced us to reembody the spectator who produces this vision.[61] The natural historian's remove and curiosity at flying fish look wildly at odds with the enslaved person whose hunger could compel him to risk his life for those very fish. And, in a view that implicates the reader as well, we might say that the natural historian can only be amazed at the fish's aerial abilities—simply amazed—if he has not seen human bodies chart that same trajectory, successful in death or brutally interrupted in their flight. Packed into this apparently superficial moment, then, is this question: can we ever observe—or even read about—nature again without feeling whether our own bodies are healthy, are hungry, are tortured, are respected?

We can see, certainly, that Equiano is not inclined to offer a positive definition of subjectivity to which the reader might simply assent. For this reason the way in which he leads us to reflect on this question may be as important as the final result: the body onboard the ship. In the cases of the flying fish, Equiano makes the definition of subjectivity a felt experience for the reader. By this I mean that he allows us to "see" as we are used to, through a natural history lens; then, through the way we unpack our surprise and understanding, he allows us to come to an understanding of embodiment. What he forces here is a kind of phenomenological experience of subjectivity. We might note the vast difference of this from sympathy, which also might be part of our interaction with the text. For here the invitation is not to borrow someone else's experience but instead to comprehend embodiment, each of us, on our own terms.[62]

Turning to Latour's formulation of the fetish, we might say that what Equiano shows us about Enlightenment is the way that, even on the level of perception, its perspective is "crafted." Latour is interested in later science, in the Moderns' denial that the objects of scientific investigation are themselves made things, not so different in this case from the pipe full of bodily materials in Bosman's account of Guinea. This is the kind of point that Equiano is making in the passage on the portrait and that, as we will see, he continues to make in other central passages on slavery in the text. A long time before the anthropologists and before Latour, then, Equiano advances the position that we are all fetishists, not only to point out analogy and sameness, but to do so by way of interrogating the epistemological assumptions on which the division between "us" and "them" has been thought to rest. This is a radical move. He makes it from the position of the fetishist, the debased object of European culture's dominant epistemology, and he makes it in order to reveal to the reader that the freedom for which antislavery advocates continually plead must be accompanied by a rethinking of how knowledge is produced. This is not something that one proposes on the floor of Parliament. But it is, as Equiano rightly perceives, a kind of work that literature is able to do.

We can already see that Equiano, in his attention to the way that perception itself is made on both sides, on all sides, is working around the individual bourgeois subject that the genre of spiritual autobiography seems to entail. Thinking back to the passage on the clock and portrait once more, we can see that Equiano calls our attention to the sort of subject that might make up this world in the first place. Here again, superstition instructs. The young boy worries that the objects tell and look, but they are not, I think, "surrogates for the master" in any simple sense.[63] That the master is sleeping makes the point that they act without his direct agency. That makes their "surrogacy" less an analytical position that rationalizes what the boy points out from superstition, more akin to the boy's own sociological musings, full of the troubled causation of which fetishists are routinely accused. But it is not just that. Equiano teaches us something about the fetish in this passage, not just by pointing out that Europeans (and anthropologists and literary critics) fetishize too, but that by accepting that logic we also have to do away with the singularity of the subject as the master himself—the slave master as master-Enlightener—envisions it. Like the power that might come from a magician or the coffin bearers, like the watches who think for their masters, the fetish works to undo our typical assumptions about agency. As Latour puts it, the fetish assaults "the very certainty that mastery is possible."[64]

While Latour finds his craft in the metaphors of constructivism, Equiano finds his craft in the very natural philosophy that took it as a central term. Like many other authors in this book, Equiano takes the craft basis of natural philosophy's account of experience and raises it to prominence, exploits it, in order to refashion the idea of experience altogether. But this is not an easy or obvious move for Equiano to make. It is rare to find an antislavery text in the period which chooses to defend African humanity, in the context of chattel slavery, on the basis of craft—rather than, say, the more common thinking mind or loving soul. Craft is an embodied knowledge. Better, antislavery writers seem to have thought, to leave aside a formulation of identity that could not but call the reader back to the world of slavery and its ownership of the body. This is the challenge that Equiano takes on, armed with the fetish.

Craft Without Mastery

If Equiano undermines the concept of mastery through describing the master on the slave plantation, he does something similar with the slave ship and the Middle Passage: offering us a counterlogic, a way of understanding how to think in ways that would permit the situation to be otherwise. This begins, we might say, not so much with what Equiano does as with what he refuses: the logic of sentimentality. The famous slave-ship scene in *The Interesting Narrative* is full of the feeling responses dictated by eighteenth-century conventions of sentimentality. Again and again, the passage records the young boy's "horror" and "fear." It shows him faint at the sight of the other enslaved Africans on the ship, important evidence that he, an enslaved person, has a body that fears and feels—in sentimental terms, he is human. Brycchan Carey has detailed the importance of sentimentality for both pro- and antislavery writers of this period. And he ranks Equiano among "the most vociferous critics of sentimental proslavery writing."[65] But when it comes to sentiment in *The Interesting Narrative* itself, Carey seems less certain about how to proceed. He acknowledges that "Equiano's narrative of his childhood invites a sympathetic response" and he points up two different "sentimental set-pieces," when Equiano uses feeling to draw characters together and to draw the reader into the text, both of them related to his sister: when he is briefly reunited with her ("Our meeting affected all who saw us," Equiano remarks, thus instructing the reader) and when, following that, she disappears forever (*IN* 35).[66]

The slave ship does not make Carey's list. But looking closer at this passage shows us an even fuller modification of the sentimental than Carey leads us to imagine. Equiano's letter to the *Public Advertiser* of 1788, which attacks Gordon Turnbull's *Apology for Negro Slavery*, declares, "were I to enumerate even my own sufferings in the West Indies, which perhaps I may one day offer to the public, the disgusting catalogue would be almost too great for belief."[67] This has been taken, by Carretta and others, as a potential advertisement for *The Interesting Narrative*. It is intriguing, then, that it is precisely his "own sufferings" that Equiano keeps at bay. His examples of tortured bodies in chapter 5 are of other enslaved persons, himself a witness. And, following this, minute descriptions of violence on the body are extremely curtailed. Indeed, reflecting on the subject matter of the "cruelties" he has offered in the fifth chapter, Equiano states clearly at the beginning of the next chapter that "The punishments of the slaves, on every trifling occasion, are so frequent, and so well known, together with the different instruments with which they are tortured, that it cannot any longer afford novelty to recite them; and they are too shocking to yield delight either to the writer or the reader. I shall therefore hereafter only mention such as incidentally befel myself in the course of my adventures" (*IN* 90).

Equiano steers clear of the bodily particulars of torture in what follows. The closest he comes to relating the tortured body lies in his account of how the slave owners on a plantation in Georgia "beat and mangled" him; but the specifics of the attack, and the particular harm to the body, are not related to the reader (*IN* 103). At the outset of his book *Spectacular Suffering*, Ramesh Mallipeddi associates Equiano with the "desire to witness and document physical punishment in the interests of legislative reform"; indeed he uses Equiano to introduce and define this concept.[68] But the textual example Mallipeddi pulls from *The Interesting Narrative* is a bit more complicated than he lets on, for it is the moment when Equiano relates that he "dreaded, of all things, the thoughts of being striped," and was "very apprehensive of a flogging" (*IN* 116). This is a flogging that does not come off, and Equiano jumps in to reassure the reader that his dread is related to the fact that "I never in my life had the marks of any violence of that kind" (*IN* 116). It may be the case that Equiano "speaks not as an observer but as a victim," insofar as he is an enslaved person and unquestionably subject to violence, but the passage ends closer to the account of Sadiya Hartman, which Mallipeddi critiques, than to Malipeddi's own.[69] For Equiano remains in tight control of how the reader witnesses violence. And he represents violence, in this case, while powerfully separating his own body from it. It says a great deal about the text, I think, if this "thought" is the most "spectacular" that its narrator's suffering gets.

Equiano's consistent separation of his own body from the violence he depicts is so extreme and pointed, and his claims about the troubles of excessively violent representations are so plain, that it seems as though he has in mind outright avoidance of the troubling potential of graphic violence on the body. Hartman, in terms which echo Equiano's earlier ones, describes the problems with the "routine display of the slave's ravaged body," which, despite its antislavery aims, participates in the "routinized violence of slavery."[70] Hartman explains the logic of these earlier texts in representing the "minutest detail of macabre acts of violence": "By bringing suffering near, the ties of sentiment are forged."[71] "In this case," she continues, "pain provides the common language of humanity; it extends humanity to the dispossessed and, in turn, remedies the indifference of the callous."[72] But, of course, this is a highly problematic thing to share and to presume to be able to share. As Hartman unpacks the problem, "Can the white witness of the spectacle of suffering affirm the materiality of black sentience only by feeling for himself? Does this not only exacerbate the idea that black sentience is inconceivable and unimaginable but, in the very act of possessing the abased and enslaved body, ultimately elide an understanding and acknowledgement of the slave's pain?"[73] As she puts it in declarative form, "Empathy is double-edged, for in making the other's suffering one's own, this suffering is occluded by the other's obliteration."[74] Hartman, writing on the later American context, uses the term *empathy*, but Equiano's own term would be *sympathy*, whose mechanisms were well acknowledged, even within the period, to reinforce the self, rather than finding a way toward the other.[75]

Where feeling goes—and who feels it, and how—are questions we see Equiano approach directly on the slave ship. Surely the scene of enslavement would be an obvious place to provide the reader with an opportunity for sympathy. When Thomas Clarkson describes the embarkation of a slave ship, for example, he describes it this way: "The unfortunate people, that have been put on board, separated from their families and friends, on the verge of bidding adieu to their native country, which they yet behold with streaming eyes, and about to depart into a servitude of which the most horrid notions are entertained, cannot but be supposed to be in a forlorn and melancholy state."[76] After this general statement, Clarkson moves to describe instruments of punishment to keep the enslaved from committing suicide. Then, he describes the plight of a particular "beautiful African girl, who had reached her sixteenth year," who does not make it past enslavement on the ship; before they can leave the coast of Africa, she commits suicide.[77] Lest the reader not respond correctly, Clarkson offers some standard sentimental rhetoric to guide feeling: "Poor unfortunate girl! What availed the care taken in her infancy to support her! The anxiety of the mother!"[78]

Equiano does describe the group of Africans he sees standing on the ship: "a multitude of black people of every description chained together" (*IN* 39). He does not avoid feeling here, but he does not individuate the sufferers and circumvents direct attention to their bodies as sites of punishment. Instead, he relates to us "their countenances expressing dejection and sorrow" (*IN* 39). By focusing on "countenances" and "expressions," even as the scene of slavery insists on bodies, Equiano abstracts these cues to sympathy, but forces us to rely on him to deliver up that material for affective connection. We might understand Equiano in this scene as a kind of protective intermediary, reading bodies for us, telling us how to understand what they say. He displays the feeling nature of his own body when he faints at the scene: "every one of their countenances expressing dejection or sorrow, I no longer doubted of my fate; and quite overpowered with horror and anguish, I fell motionless on the deck and fainted" (*IN* 39). As we think back to Hartman's example of the white reader who takes the Black body's suffering as his own, we can see here Equiano's countermaneuver, for it is a Black body who faints at the suffering of Black bodies—a closed circuit to which the reader can only bear witness.

Despite the care that Equiano takes in parceling out feeling and in specifying the conditions for it, scholars have often read right through these efforts and have reasserted the model of sympathy as a means of communicating pain. When Marcus Rediker, in his book on the slave ship, retells the punishment inflicted on Equiano for refusing to eat, he puts it this way: "When two members of the crew offered food, he weakly refused. They hauled him back up to the main deck, tied him to the windlass, and flogged him. As the pain coursed through his small body, his first thought was to try to escape by flying over the side of the ship, even though he could not swim."[79] The historian's dramatic paraphrase diverges substantially from the way Equiano describes the incident:

> two of the white men offered me eatables; and, on my refusing to eat, one of them held me fast by the hands, and laid me across I think the windlass, and tied my feet, while the other flogged me severely. I had never experienced any thing of this kind before; and although, not being used to the water, I naturally feared that element the first time I saw it, yet nevertheless, could I have got over the nettings, I would have jumped over the side, but I could not; and, besides, the crew used to watch us very closely who were not chained down to the decks, lest we should leap into the water. (*IN* 39–40)

As is obvious in a quick comparison of the two passages, Rediker's "the pain coursed through his small body" is entirely the later author's own. It is meant, perhaps, as a correction to Equiano's own failure to mention pain or to describe his material body. Here, in a long tradition of antislavery writing, Rediker gives us the materials for salvific and problematic sympathy, pushing the body to the fore, offering "pain" as the universal language Hartman described.

Reliant on this model of pain and sympathy to construct humanity, and concerned with rereading along these lines, Rediker also misses the objects that share the scene of suffering, given to us through terms like "windlass" and "nettings." The narrator's preoccupation with such objects is evident in the strange piece of information he gives us: he was flogged, he says, while "laid . . . across, I think, the windlass"—where the hesitation directs us to focus on "windlass" rather than on the beating. In later passages, when Equiano becomes accustomed to the mariner's life and learns the ways of the ship, in the sea adventure sections, certainly, such terms are right at home, and the modern editorial glosses from William Falconer's much read *Universal*

Dictionary of the Marine (1780) become thick.[80] But in the reader's encounter with the slave ship, in all of its horror and terror, the maritime references feel out of place. What is the role of these specialized tools, and what relation do they have to the violence on the slave's body that takes place on or around them?

The maritime terms are hard to account for in the scene of enslavement: surely the part of the ship over which the enslaved person is flogged is irrelevant, the windlass merely a surface on which the violence to the body occurs. In this they have something in common with the "scandalous" details of Barthes's "The Reality Effect": Flaubert's "barometer" and Michelet's "little door."[81] In Equiano, too, maritime details seem "superfluous" and "isolated[]," separate from the plot that propels the young boy into slavery.[82] But of course in the worlds of Gustave Flaubert and Jules Michelet, society is intact, and the roles for objects are stable. Equiano's windlass, in this way, might be closer to the objects in the scenes of torture that Elaine Scarry describes; there, everyday objects—bathtubs, file cabinets, refrigerator doors—are made into weapons. As she puts it, "The room, both in its structure and its content, is converted into a weapon, deconverted, undone. Made to participate in the annihilation of the prisoners . . . the objects themselves, and with them the fact of civilization, are annihilated."[83] Here, then, the object that doesn't fit does not gesture toward the reality of the scene, its excess a testament to the referential program of the realism at hand; no, rather, it does not fit because it is the weapon taking apart that reality, its excess destructive of what we thought we knew.

Although I think that Scarry's description of weaponization is accurate to the passage, Equiano's account does not end with the undoing of civilization that Scarry describes. To begin with, the windlass is not quite the door or cabinet of Scarry's torture scenes. It is a routine object but only in what Rediker refers to as the "wooden world," the world of the ship.[84] In this way Scarry's analysis points us to Equiano's focus on a very specific reality that is being undone: the reality of seafaring life. The re-reader of the text in particular, however, experiences the tool as double-edged: weaponized and undoing, it also recalls for the reader Equiano's later maritime scenes of freedom. And in this sense, it might be understood to take apart but also to rebuild. Indeed, it calls out to Equiano's first employment as a free man: his work as an "able-bodied sailor" (*IN* 115). In this sense, the windlass is an especially concentrated instance of what Equiano has encouraged us to see all along:

within the slave ship is the ship itself. In the beginning of the narrative, ships are associated both with enslavement and with a physical way out; even when he describes Africans being taken below decks at the slave ship's departure, Equiano notes that this deprives them of knowledge of the ship's workings. Later, following this logic, the captain of the *Nancy* prevents Equiano from learning navigation, sensing his loyalty but fearing his escape. Again and again it is clear that the very vessel containing enslaved persons could also offer those persons a way out.

In this sense, the windlass is at once a kind of weapon and a portal to other scenes on other ships. I think we can understand this moment as working similarly to that on the plantation, where Equiano takes a pivotal moment in his own history of slavery—his first view of the master—and embeds within it a way out: not a way out of bondage at that moment but a way out of the thinking on which that bondage relies. It has both the touch of the political radical and also that of the careful historian who is aware of the way that his own constructions will enable our views of what is possible next. I think here of W. Jeffrey Bolster's moving history of Black seamen, where he asks us at the outset to consider the way that history's focus on the Middle Passage has made for only one kind of association of Black men with ships; this immediate connection to slavery, he says, "reinforces whites' belief that blacks were acted on, rather than acting; that blacks aboard ship sailed as commodities rather than seamen."[85] To use Bolster's terms in returning to Equiano, we might say that Equiano, as an earlier historian of slavery, stages the scene of enslavement, the most violent ways of being "acted upon," in discourse that opens up to acting.[86]

The "reality" that Equiano thus insists can be recovered is a maritime reality, and in that world the windlass has a very particular function, one that can give us insight into what kind of an actor he wishes us to envision for the postslavery future. As Falconer has it, to operate the windlass men turned in a circle and the machine "requires, however, some dexterity and address to manage the handspec to the greatest advantage; and to perform this the sailors must all rise at once upon the windlass, and, fixing their bars therein, give a sudden jerk at the same instant, in which moment they are regulated by a sort of song or howl pronounced by one of their number."[87] As Rediker puts it succinctly in his account of this passage, "Seafaring labor, in its work chants and songs, revealed its profoundly collective nature."[88] On Rediker's account, then, the windlass is a kind of emblem for what he finds in maritime world more generally. The "wooden world," separate from the societies on shore, is home to "a densely communal life"; it has its own "culture": "conventions, symbols, and rituals."[89]

We can see, just within the detail of the windlass, then, something of what Equiano wishes to revise in the models of subjectivity that are available to him. The odd locution "across, I think, the windlass," points to the tool in terms of thought. In his hesitation, his lingering over the windlass, Equiano draws up the tool as a product of his memory. This way of drawing in might seem to indicate a subjectivity predicated on the form of consciousness-through-autobiography long associated with this text and, more broadly, with Ian Watt's "formal realism" and the novel. But if Equiano gestures toward the consciousness we would expect in this way, he at the same moment looks beyond the insular, single individual person that it would seem to entail. On the one hand, the individual's memory-object contains himself as a body dehumanized, less than a subject; on the other hand, it contains the group, the collective subjectivity of the sailors. These forms of subjectivity aren't so much reconciled by the commanding subjectivity of the writer as they are nested, one within the other.

In this account of collective subjectivity Equiano begins to perform some of his most startling reversals of the logic of mastery. As an elegant writer of English who can quote from *Paradise Lost*, Equiano would seem to attempt a text that succeeds because it masters the master's own tongue and endorses its ideas of cultural capital. For this very reason Equiano's most skeptical readers in the eighteenth century required Equiano to defend himself as the true author of the text; later critics, though no longer doubting authorship, have turned to the same textual cues to marshal criticism of the work's capitulation to white, European standards. Equiano no doubt had a great deal at stake in his own cultural advancement as proof against slavery. His portrait printed at the front of the volume speaks to his elevated cultural identity, his distance from the limited intellectual capacities assumed for Blacks. In a world where Phillis Wheatley had to explain her poetry to a room of powerful white men in Philadelphia, we can see that Equiano has little choice but to take this tack: without it, the text might not be read in the first place.

But when we turn to the positive forms of subjectivity that Equiano raises within the text, we find both the most radical political project of *The*

Interesting Narrative and, again, craft. As the windlass suggests, and as later passages will bear out, Equiano's craft here is that of the mariner. Indeed in this sense he follows Margaret Cohen's formulation of the sailor as a definitive Enlightenment subject.[90] Navigation, a skill central to Equiano's own depiction of his seafaring education, is an old form of craft knowledge, an example of Aristotle's practical knowledge and a convenient example of applied knowledge ever since. It is also an embodied knowledge, even if it might sound at first like the more cerebral counterpart to routine maritime work. Navigation, as we will see, is not separate from these other tasks, either in its individual actions or in kind.

Cohen explains this in terms of the role of the captain, focusing particularly on the Barrier Reef passage from Captain James Cook's journal of the *Endeavour* voyage. Cohen calls Cook's "arguably the most famous remarkable deed in the annals of navigation," and she finds it exemplary of the mariner's "craft."[91] Such craft, which highlights the "embodied reason" that Cohen also refers to as "practical reason," demands a narrative that focuses on action, rather than on psychology.[92] And it demands a hero whose intelligence is as much bodily as contemplative. She shows, for instance, how Cook excels at both "care and foresight" (prudence), even as he demonstrates his "sea legs" by "personal involvement" in both assessing and remedying the situation—a "bodily participation" which "demonstrates the importance of physical strength and agility."[93] In the rescue he orchestrates of the ship and crew, Cook reveals endeavor, or patience, as well as resolution, ultimately "turning hazard into salvation."[94] And he offers this salvation through what Cohen terms "jury-rigging," a form of "creative improvisation" most clearly demonstrated by the fothering he performs in repairing the ship, detailed through a recipe involving oakem, wool, and dung.[95] The "compleat" mariner, while part of a collective, also does it all: from the high-powered navigational duties with attendant mathematical calculations we would associate with those highest in the ship's hierarchy, to a more physical examination and repair of the ship, which we might assume were lower duties given exclusively to the ordinary sailors.

Equiano stresses just the aspects of maritime work that Cohen does here, but he additionally calls attention to a malleable hierarchy we recognize from Rediker. This is presented in the narrative at a critical moment: in November 1766, in Equiano's first employment as a free man, he enrolls as "an able-bodied sailor, at thirty-six shillings per month" aboard the *Nancy*, sailing for St. Eustatia (*IN* 115). Almost immediately, Equiano is physically assaulted while onshore and his free status is questioned. He is ultimately returned to the ship (with the intervention of the captain). But we are shortly to see how very differently social hierarchy on the ship works for the recently free man. For when the ship directs its course toward Monserrat, the captain and mate of the ship both take ill, and Equiano is left with the "whole care of the vessel" during a major storm (*IN* 119). Both of the designated skilled navigators on the ship, then, are laid low. Finally in a kind of wild literalization of Cohen's description of "practical reason" for the sailor, Equiano declares flat out that, unable to use a traverse, he "was obliged to direct [the ship] by mere dint of reason" (*IN* 119). He not only uses this unaided "practical reason" to sail through the storm but, on the captain's death, pilots the ship into port, where he says, "I now obtained a new appellation, and was called Captain" (*IN* 120). Equiano clearly takes enormous pride in this designation, but that should not obscure for us what it shows us about how power and authority on the ship can work. Equiano's triumph is exceptional but also relies on something particular about the experienced-based knowledge present on ships, the fact that the hierarchy of command did not always reflect the distribution of knowledge among the sailors. It was frequently the case that an "able-bodied sailor" (defined as such precisely through his experience with other voyages) of Equiano's ilk might have experience equal to—or in some cases in excess of—the captain or mate. Thus, even though Black men were not permitted to command merchant ships, this seeming absolute could be, and in fact was, very much undone, rendered irrelevant in actual day-to-day situations at sea. As Equiano demonstrates here, due to their command of maritime knowledge common sailors could step into the top job. And it was possible to do so regardless of the color of one's skin.

Having shown us his ability to command, and to do it through the practical reason of an excellent captain, Equiano has set the stage for what happens next. Here, again on the *Nancy*, in a subsequent voyage to Georgia, in January of the following year, Equiano relates his own struggle with a reef, in a passage that very closely resembles Cook's. (It is possible that Equiano knew Cook's passage from Hawkesworth's edition of the journals from the voyage.) Here too, Equiano proves the prudent and active seaman who can easily seize control of the ship and guide its passengers to safety.

222   Chapter Six

The episode begins with Equiano's dreams of a shipwreck and is structured around providential appeals, in a way that the mariner's earlier success during the storm is not. Even with these markers of spiritual autobiography, however, what comes through again are Equiano's extraordinary practical abilities and demonstration of expertise.[96] At first the "man at the helm" calls Equiano up to see what he believes is a grampus, an object Equiano then identifies as "not a fish but a rock" (*IN* 122). At this point, he warns the captain of the danger; the captain neglects the prudent and patient warning twice and only on Equiano's "growing quite enraged" and asking a third time does he come up, too late to avert the danger (*IN* 122). Within the text of the description, Equiano at this point already seizes control of the situation and the ship, collapsing in the first person plural his own activities with the captain's : "We then called all hands up immediately; . . . we got up one end of a cable, and fastened it to the anchor" until "we let the anchor go" (*IN* 122). With the ship wrecked on the rocks, "while the dreadful surfs were dashing with unremitting fury among the rocks," Equiano calls on God (*IN* 123). He also stresses his own active mind, that creative capacity so important for the seaman: "I believe no mind was ever like mine so replete with inventions and confused with schemes," he tells us (*IN* 123). Finally, Equiano takes control of the ship, and he admonishes the captain for his lack of seamanship: "I told him he deserved drowning for not knowing how to navigate the vessel" (*IN* 123). Equiano then assumes control, asking the crew to "get the boat prepared against morning;" while the captain and white crewmen abandon themselves to drinking, Equiano jury-rigs a solution for the hole in the ship, reminiscent of Cook's solution to the *Endeavour*'s hole (*IN* 123). As Equiano relates it, similarly stressing the practical through the ingredients necessary for the task, "I took some pump leather and nailed it to the broken part, and plastered it over with tallow-grease" (*IN* 124). Following this, again acting without the help of the white sailors, he works with other sailors to "drag" boats full of people and supplies over the reefs and onto a small island (*IN* 124). Equiano is master of both plan and execution, "inventing" in the face of "necessity" and toiling with the others when toiling is called for (*IN* 124).

Equiano draws attention to this episode as an important one for the text as a whole: there are only two illustrations in the first edition of the text and one is of the wreck of the *Nancy*.[97] All of this action and practical reason convinces us that Equiano is the "compleat mariner." But the passage contains a moment that could never appear in Cook's account of the *Endeavour* voyage, for Equiano is a formerly enslaved man aboard a slave ship. Thus, the capable attempts to save the ship are interrupted by the bodies in the hold:

> The captain immediately ordered the hatches to be nailed down on the slaves in the hold, where there were above twenty, all of whom must unavoidably have perished if he had been obeyed. When he desired the men to nail down the hatches I thought that my sin was the cause of this, and that God would charge me with these people's blood. This thought rushed upon my mind that instant with such violence, that it quite overpowered me, and I fainted. I recovered just as the people were about to nail down the hatches; perceiving which, I desired them to stop. The captain then said it must be done: I asked him why? He said that every one would endeavour to get into the boat, which was but small, and thereby we should be drowned, for it would not have carried above ten at most. (*IN* 123)

Even if we don't have Cook's passage in mind, this is a startling moment in the narrative, something of a surprise to the reader who is at this point fully in tune with a familiar sort of heroism.[98] Equiano is at peak mental and physical performance but then, for a moment "overpowered," he loses control of both mind and body. Moreover, this moment repeats the one on the first slave ship, when Equiano was a boy, newly enslaved. Through this connection to other Black bodies, we are reminded of Equiano's own status, his history, his body. But then, as quickly as the faint occurs, it is over. The passage barrels ahead with action. Equiano loses nothing and gains his aim: he rises to question the captain directly; the hatches are not nailed down.

Equiano's reflections and his demands of the captain in this passage cast this as a story of salvation. When God tells Equiano that he will have the blood of the enslaved on his hands, he also positions Equiano as a Moses figure who could lead his people from bondage. But the readers of Equiano's autobiography quickly can see that this is no simple act. To begin with, it is staged within a moment when Equiano is himself complicit with the system of slavery; after all, he is not enslaved onboard this ship but a slaver, a free Black man likely charged with managing the roughly twenty enslaved persons aboard ship. We must begin by noting, then, the way that the later antislavery writer Equiano goes back and rewrites a scene of salvation from within his own complicity with chattel slavery more generally. The passage shows in miniature what Carretta has said about *The Interesting Narrative* as a whole: it is an attempt for Equiano himself to come to terms with his own changing views of slavery.⁹⁹ This particular case should call our attention to Equiano's larger project and particularly to his engagement with the idea of "heroism" itself. We might say that Equiano's revisionism is not merely about alleviating his own guilt over bad actions, but indeed about showing us how to think about good, ethical actions.

The captain of the *Nancy* in this scene is not just any bad captain. Equiano's language in the scene draws a connection to Luke Collingwood, captain of the *Zong* and one of the people chiefly responsible for the massacre of enslaved persons on that ship in 1781. As is well documented, Equiano called Granville Sharp's attention to the massacre, one of the most horrific tragedies in the late-century slave trade and one that was made famous by the antislavery cause.¹⁰⁰ We might note the similarity between the white captain's kind of thinking—"every one would endeavor to get into the boat, which was small, and thereby we should be drowned, for it carried ten at most"—and that claimed by the officers of the *Zong* who had drowned enslaved people by throwing them overboard due to a lack of water. Both include the same kind of calculus, the counting of bodies that could survive, the subtraction of slaves deemed less than crew.

If this is the man Equiano bests in seamanship and action, however, Equiano represents his own action in a way that is downright shocking given Ian Baucom's analysis of the *Zong*.¹⁰¹ Although it is the enslaved persons below deck that motivate Equiano's identity as a savior, the remainder of the passage—following the hatches that are not nailed down—neglects their description entirely. Equiano tells us that he and a small group of men "went with the boat five times that day," and that the white men were so drunk his group had to "lift them into the boat, and carry them on shore by force"; he says nothing of the persons who were once below deck (*IN* 151). He concludes this way, "out of thirty-two people we lost not one" (*IN* 151). To offer this conclusion as enumeration seems dangerously close to the logic that Baucom associates with slavery, to the "abstract form" that debases persons and turns them into "types" or numbers.¹⁰² Indeed when Baucom reads closely the log of another slave ship contemporary with the *Zong*, he draws attention precisely to the relation between the person's existence only as a "type" and the captain's failure to record any of the particulars of the men and women subject to enslavement.[103] To let the particulars fall out of representation comes, in Baucom's analysis, with a heavy ethical charge. Given Equiano's disparaging depiction of the captain, we would expect him to refuse this logic of representation.

Yet it is not clear at all that Equiano shares Baucom's paranoia about the formal act of abstraction itself, what Baucom calls "actuarial realism."[104] And indeed, we can read the larger passage—of which "thirty-two" is a part—to suggest a different role for the number-as-abstraction. If Baucom's argument would convince us that a more humane representation in this scene might have particular enslaved persons named and described, Equiano offers us an account of personhood that proceeds otherwise. It begins with God as the one who suggests that the deaths of "the slaves in the hold" might be Equiano's responsibility: he thought, he tells us, "that God would charge me with these people's blood." God, then, makes the first bid for personhood for the slaves in the hold: they are, in this moment of terror for the narrator, "people." Equiano nearly repeats this phrase three paragraphs later. When the drunken white sailors refuse to leave the ship, Equiano again reflects, "I could not help thinking, that if any of these people had been lost, God would charge me with their lives" (*IN* 149, 151). "These people" and "these [other] people" come together in the total number saved from the ship. In this sense, we might say, Equiano takes advantage of abstraction to smuggle in difference; what begins as slaves and sailors ends as "people" saved, each counted as the same in kind.

The passage continues to instruct us, not only on the persons saved, but on how we should conceive of the saving and the savior attached to them. We recall, for instance, that when Equiano berates the captain he does not voice God's language of salvation. Equiano does not accuse the captain of inhumanity, exactly, does not put in his own mouth some version of the fear God gives him. Rather, he accuses the captain of bad seamanship. And Equiano's own solutions follow in these terms: his triumph as a savior overlays the Judeo-Christian savior—Moses or Christ—with the giver of "salvation" Cohen finds in Cook's timely and practical actions.[105]

Yet when Equiano turns to the person of the "compleat mariner" as a kind of secular savior, it is with a difference. Cohen, when she describes the mariner, seems to assume that practical knowledge is practical knowledge—regardless of the body involved in its embodiment. Equiano, however, goes out of his way to show us that this is not the case in the way he mobilizes collective identity. After Equiano calls out the captain's poor skill, he remarks, "I believe the people would have tossed him overboard if I had given them the least hint of it" (*IN* 150). In this aside Equiano overlaps his role as captain that of another powerful figure: the leader of a mutiny. Equiano positions himself alongside the captain with the "we" of leadership; on the other hand, he signals a power that comes from outside that hierarchy and is in fact entirely hostile to it. As Rediker has described them, mutinies play a very powerful role in the history of collective action. In a mutiny, "Sailors then expropriated the workplace and arranged it anew," as he puts it.[106] And, further, "Given, in fact, the logic of collectivism that informed seafaring work, it comes as no surprise that the very term 'strike' evolved from the decision of British seamen in 1768 to 'strike' the sails of their vessels and thereby to cripple the commerce of the empire's capital city."[107] Equiano's passage also recalls slave insurrection, pulling together two kinds of uprisings that Rediker and Peter Linebaugh link together in *The Many-Headed Hydra*.[108] In the above passage Equiano has it both ways, leading and leading an attack on the leader. In doing so, he remakes the model that Cook's passage puts forward, the collectivity that still relies on the "absolute power" and mastery of the captain (or the aptly titled "master"). Instead Equiano asks us to imagine a form of leadership that undoes the relation of the master to the collective. He continues this in performing an act of salvation through working with a group of other sailors who are either "black" or "creole." "There were only four people that would work with me at the oars; and they consisted of three black men and a Dutch creole sailor; and though we went with the boat five times that day, we had no others to assist us" (*IN* 151). Thus, although Equiano describes himself at the top of the hierarchy and in terms of exceptionality—"I could not help looking on myself as the principal instrument in effecting our deliverance"— this is cut through with the many repetitions of "we," with the particular composition of a group that works together in order to save (*IN* 151).

The passage is, then, consonant with what we know of Equiano after his writing of *The Interesting Narrative*. In *The London Hanged*, Peter Linebaugh describes Equiano as a "leading London activist," who "in early 1792 . . . helped Thomas Hardy link up with the Sheffield abolitionists and the Corresponding Society, thus extending the constitutional discussion of reform from plebian to proletarian contexts nationwide."[109] John Bugg's more recent work describes the means by which Equiano attempted a "project of abolition from below," bringing his book tour to the "industrial north" with its "world of miners, glovers, and grocers, calling on a community of readers who, he hoped, would realize the transition from literary sympathy to political activism."[110]

In this reaching out and sharing of cause, it is important, then, to note *The Interesting Narrative*'s conception of the sailor's identity, particularly in terms of the team of Black and creole sailors who call our attention to race as an aspect of embodiment. If much of the seafaring scenes of *The Interesting Narrative* is about asserting that a Black body can function the very same way as a white one does—a sailor's experience is a sailor's experience—this act of the small group of men pointedly described as other than white, turns a philosophical account of experience another way. Here we have the pointed insistence on the body that we found in the scene of the young boy peering through the quadrant. We are given a Christian narrative of salvation, but rather than the individual white savior we might expect, we have Black and creole men instead. Moreover, the work of these men pointedly recalls that of slavery: dragging the other sailors to shore was, he says, "labour intolerably severe" (*IN* 151). Equiano continues to insist on the particularity of his body when he reports that, as a result of the work, "the skin was partly stript off my hands" (*IN* 151). Equiano is so careful with his own self-presentation in *The Interesting Narrative* that this is the most graphically particular of any violence to his own body.[111] The relation of the "stript" hands to his body which was not "stript" can hardly be an accident. Equiano chooses here to report broken flesh, to recall for us slave violence, to claim some version of it for his own body. We might think that at this moment of heroism, when he manifests as a free man able to do a white leader's work better than a white man can, Equiano would shirk from a reference to slave violence. But to assume this would be to assume that Equiano ultimately wishes to embrace a universal subjectivity, embodied or not. We see him resist this again and again, showing us that embodiment for a Black body cannot escape the violence of slavery, even as he shows us too how that violence need not result in victimhood but can mark a body that is willful and strong.

I recall here Hartman's disparaging account of the "heroic actor of the romance of resistance," the way in which that account of heroism is something the enslaved person, acting at all times under crushing levels of subjection, cannot possibly attempt; as a result we later readers of the acts of the enslaved can find ourselves unable to conceive of their resistance.[112] In his attention to the minute actions of the material bodies of the group that saves, and in his attention to craftsmanship as an aspect of subjectivity, Equiano insists on a model of the subject that does not transcend the material body and indeed that must work within its past and continuing state of oppression. Hartman, that is, allows us to see how this passage goes two opposite ways on "heroism," with Equiano claiming his role as the "principal instrument" and, at the same time, forcing us to see this absolute individual heroism as both collective, and violently embodied. Hartman stresses the inability of the body to transcend the violence done to it, and we can read this pessimistic strain in Equiano's account, as well.

Hartman directs us to return to de Certeau's "practice" as a way of conceiving of resistance, and the turn toward practical knowledge is a part of Equiano's ethical account, too.[113] For when Equiano accuses the captain of bad seamanship (of navigating poorly), this carries its own ethical charge. In Equiano's modification of Cook's model, he refashions the heroism but he keeps the craft knowledge, one attached to process, problem solving, decision making. Richard Sennett suggests to us why this might be when he positions the craftsman over and against Hannah Arendt's division between *animal laborens* and *homo faber*. For Arendt, *animal laborens* is "the human being akin to a beast of burden . . . absorbed in a task that shuts out the word."[114] By contrast, in Sennett's formulation, *homo faber* "is her image of men and women doing another kind of work, making a life in common."[115] For Arendt, this is a political making, a "higher way of life in which we stop producing and start discussing and judging together."[116]

We recognize in Arendt's separation a twentieth-century incarnation of the old division between theory and practice with which this book has been concerned. Sennett's reading of Arendt—his suggestion that the craftsman undoes this opposition between making things and making a common life—helps us to see the importance of craft for Equiano's antislavery project. Making can be salvific, but it does not transcend the world from which it originates, and it does not transcend the bodies of its actors. It is no less, and perhaps even more, transformative for this. Equiano allows us to see that the rebuilding of life proposed by the world of the sailor, layered within the experience of the enslaved boy on the slave ship, is emancipatory in no simple way. Equiano gives us as a way to think about a world that is not driven by slavery. He offers us, at once meager and everything, an example of how things can be remade, can *work* differently—and without mastery.

CODA

Handicraft Humanities?

HANDICRAFT PHILOSOPHY, IT SEEMS, IS alive and well in the twenty-first century. At a time of crisis in the humanities, critics have begun to reevaluate the kind of work we do, most famously in the so-called "method wars" around "critique." As Rita Felski puts it in *The Limits of Critique,* we should rethink reading, not as "an unraveling of meaning," but rather as "a form of making." In the constant institutional and cultural disputes over the worthiness of the humanities, it surely is much harder to defend the gloomy work of critique's takedowns than a creative work of building up or "assembling."[1]

Other critics have heard the call for a move toward making. Patrick Jagoda advocates for a "critical making" that could help us to reconfigure our ideas of theory and practice.[2] David Staley suggests that we join the "maker's turn," creating "material forms" that still may function as "scholarship." Surely, he says, at this vexed time in humanistic study, we might move toward "hermeneutic acts" as "visual, tactile, and material objects."[3] Finally, Jonathan Kramnick uses *craft* and *handicraft* to define method for literary critics and to illuminate what he calls "the ordinary science of the discipline."[4] Focusing on close reading as "practice," he defends literary studies as a discipline by considering our practice of close reading as proximate to the work of carpenters and weavers.[5]

These diverse responses share something important: at a moment of a crisis in the humanities, critics turn toward the maker and claim maker's knowledge as their own. What can the Enlightenment context of this book tell us about this move?

The Enlightenment writers of this study allow us to see the current maker's turn as a repetition, one in which old terms appear new again in the effort to tie hand to head. Jagoda is perhaps most obvious here: his "critical making," he explains, means to resolve the binary of "theory and practice" which "has been particularly entrenched in literary criticism."[6] We recall Sprat's hybrid thinker-doers, with their "mechanical" hands and "philosophical" heads. And we recall too the way Royal Society craft was able to reinvigorate theory through this attachment, the metaphor itself demonstrating how far theory could reach into—indeed, how much it was already of—the world. In our own world where academic knowledge of all kinds is under siege, we could do worse than to reenvision our own practice as productive and part of the world in which we all live.

What might an academic embrace of craft knowledge entail? As this book has argued, craft has a hard time staying within the confines of the arguments of the philosophers who use it to reconfigure their own practice. Writers from Defoe, to Collier, to Equiano make clear how broad, how open craft epistemology is. It can describe poetic composition, theft, and a form of contingent emancipation. It can put knowledge in a lot of hands and a lot of kinds of hands; it can make philosophers of diverse nonelite persons, as this book has laid it out. I imagine this is the kind of thing Kramnick has in mind when he describes his craft methodology for literary studies as "democratic."[7] But craft might be more radically democratic than any defender of the academic humanities would desire. It operates by allowing us to see relations between what the world sets out as wildly divergent forms of knowledge and thus it might get us back to thinking about literature as a "useful art." But if it does that, it's unlikely to preserve the knowledge boundaries that the academic world sets up for its specialized efforts. Craft asks us to consider why knowledge should have this home, what its relation to other extra-academic knowledge is (including that of people who make a living from crafts used largely metaphorically in these accounts). These conversations are worth having, and here are some of the questions with which they might start. Where does our own knowledge making fit in the cultural hierarchy, and do we want it to remain there? What is the relationship between practitioners of humanistic knowledge and those who write about it, or make sculpture about it, weaving their complex sentences together or making their material objects? Craft—even on a metaphorical level—leads out toward the world, toward knowledge as more complex than we thought it was (especially if we're the ones accused of "theory"). It leads to seeing knowledge beyond the individual person, to trafficking in knowledge spaces that might be foreign to where we believe our own knowledge practices belong. In other words, it leads to questioning the theory/practice divisions set up in and by our own academic environments.

Writers of this book demonstrate well that craft knowledge does not come with a built-in ethical program, that it may work with imperialism or against it. Sometimes phenomenological or neophenomenological accounts can confuse this point, not asking us to decide why we might feel better about our work as weaving rather than just plain typing.[8] Although handicraft can suggest, as in Walter Benjamin, a rescue from capitalism, we'd want to consider how this sits with our "handcrafted" Starbucks drinks or the fact that I just got my car repaired at a place called Artisan Autoworks (before eating a slice of artisan bread). How could we pry the things we like about handicraft out of that world and discourse? Can we? What is the worth of craft epistemology, if we divest if of its (borrowed) ethical implications?[9]

And what, after all, does handicraft have to say about just plain work? Jagoda gestures toward this in passing, and Kramnick seems concerned with the academic job market. But craft asks us to think more capaciously. In this book Mary Collier mobilizes this aspect of craft, its definition as "skilled work," to her own ends, arguing that washing clothes should be considered the work of an artisan. Craft crosses boundaries in ways that ask us to reflect on the status of our own work, in all of its manifestations, beyond the production of the craft objects that are at the heart of individual arguments. What about the many aspects of our jobs that don't correspond to craft? Or does everything (teaching, service, etc.) correspond to craft, in which case we may have lost the specificity of what we're up to?[10]

Handicraft Philosophies argues that a philosophical deployment of craft provokes unsettling questions in the Enlightenment, and this is no less true in our current context. My suggestion is that we take Collier's and Equiano's lead and consider craft less a resolution to our problems than a starting point for asking about relations between hand and head and about how we value— and lead others to value—the knowledge that we make.

Notes

Introduction

1. *The Spectator*, ed. Donald F. Bond (Oxford: Oxford University Press, 1965) 1:2. Hereafter cited as *S*. In no. 10, Addison describes the project of the papers as to bring "Philosophy out of Closets and Libraries, Schools and Colleges, to dwell in Clubs and Assemblies, at Tea-Tables, and in Coffee-Houses" (*S* 1:44). Addison's terms echo Thomas Sprat's description of the Royal Society as correcting the ancients "by bringing Philosophy down again to mens sight, and practice, from where it was flown away so high" (*The History of the Royal-Society of London: For the Improving of Natural Knowledge* [London, 1667] 119). Hereafter cited in the text as *H*.

2. Addison's "Speculative Stateman" borrows from Thomas Shadwell's play, *The Virtuoso* (1676), and we can read Mr. Spectator as a version of the Gimcrack type. We are introduced to Gimcrack in his study, lying on his desk and imitating the movements of a frog in a bowl of water next to him. Gimcrack explains, "I content myself with the speculative part of swimming; I care not for the practic," thus making a hard claim for "knowledge" over and against "use" (*The Virtuoso: A Comedy*, ed. Marjorie Hope Nicolson and David Stuart Rodes [Lincoln, University of Nebraska, 1966] II.ii.84–85). As Tita Chico explains, this satire reveals the absurdity of experimental knowledge: Gimcrack performs an experiment while oblivious to "actual observable phenomena" (*The Experimental Imagination: Literary Knowledge and Science in the British Enlightenment* [Stanford, CA: Stanford University Press, 2018] 50). In comparison to Mr. Spectator's example, we can see something further: Gimcrack's example stresses an overly present body (could he not take notes on the frog?), while Mr. Spectator's example highlights the body's remove. Both are examples of the failure of Baconian science to bring together "theory" and "practice" as it claims it will. In moving from physical imitation to bodily remove, Addison adapts the model to his own purpose, highlighting the problems theory/practice divide might pose when directed at the social sphere.

3. Steven Shapin and Simon Schaffer describe this attitude as "modesty" in *Leviathan and the Air-Pump: Hobbes, Boyle, and the Experimental Life* (Princeton, NJ: Princeton University Press, 1985) 65–69. Addison's satirical approach reveals terms strikingly similar to Donna Haraway's in her critique and extension of Shapin and Schaffer. She describes how the male "modest witness" must occupy the "unmarked category," achieving a kind of "self-invisibility." Addison—in listing the social construction of the observer, as well as the bodiliness he achieves, seems well in tune with what Haraway calls "the apparatus of production of what could count as knowledge," its status as "man-made." Her term "situated knowledges" is one way to think about what the artisans of this project theorize (Donna Haraway, *Modest_Witness@Second.Millennium.FemaleMan_meets_OncoMouse: Feminism and Technoscience* [New York: Routledge, 2018] 23–25).

4. Joanna Picciotto uses Mr. Spectator as an exemplar of natural philosophical observation taken to the social sphere: an example of the professional observer whose "*seeing through*" requires a "corporeal transcendence." Picciotto, however, reads this as the price of doing business, rather than as a problem to be solved (*Labors of Innocence in Early Modern England* [Cambridge, MA: Harvard University Press, 2010] 509).

5. Francis Bacon, *Of the Advancement of Learning* (ed. Stephen Jay Gould [New York: Random House, 2001]).

6. The *Encyclopédie* and its monumental plates on artisanal making have made the French Enlightenment seem like a more obvious location than the British Enlightenment for questions surrounding the artisan. Paula Bertucci goes well beyond the *Encyclopédie* in *Artisanal Enlightenment*, challenging us to think about the period in terms of the *artiste* (an intellectually able artisan) rather than the *philosophe*, and she argues that the more practical knower is exemplary of the political project of eighteenth-century France's "economic and imperial development." Her suggestion of a general reorientation, away from philosophy and toward other kinds of knowledge making pushes in the same direction I wish to here (*Artisanal Enlightenment: Science and the Mechanical Arts in Old Regime France* [New Haven, CT: Yale University Press, 2016] 3). For another contribution to the history of Enlightenment craft, see Lauren R. Cannady and Jennifer Ferng, eds., *Crafting Enlightenment: Artisanal Histories and Transnational Networks* (Oxford: Oxford University Press, 2021).

7. Peter T. Manicas explains that the disciplinary formation of the social sciences occurs a full century later, at the end of the nineteenth century (*A History and Philosophy of the Social Sciences* [Oxford: Basil Blackwell, 1987]). See also Christopher Fox, Robert Wokler, and Roy Porter, eds., *Inventing Human Science: Eighteenth-Century Domains* (Berkeley: University of California Press, 1995), esp. the introduction.

8. Thomas Hobbes, *Leviathan* (Oxford: Oxford University Press, 1929) 160. Montesquieu, *The Spirit of the Laws* (W. B. Allen, ed., *Montesquieu's "The Spirit of the Laws": A Critical Edition* [New York: Anthem, 2024]).

9. In addition to the above, see Sergio Moravia, "The Enlightenment and the Sciences of Man," *History of Science* 18.4 (1980): 258.

10. David Hume, *A Treatise of Human Nature* (ed. P. H. Nidditch [Oxford: Oxford University Press, 1978]) 50.

11. John Smith *The Printer's Grammar* (London, 1755); Edward Bysshe, *The Art of English Poetry* (London, 1702); Israel Aber, *The Art of Manufacturing Saltpetre* (London, 1765);

John Keys, *The Practical Bee-Master* (London, 1780). See Celina Fox, *The Arts of Industry in the Age of Enlightenment* (New Haven, CT: Yale University Press, 2009) chap. 5.

12. Gilbert Ryle, "Knowing How and Knowing That: The Presidential Address," *Proceedings of the Aristotelian Society* n.s. 46 (1945–46): 1–16. Ryle separates "knowing that something is the case" (glossed as a "discovery of facts") and "knowing how to do things" (2). He aims to thicken our understanding of the second category of "practical activities" as an "exercise of intelligence" (1).

13. Samuel Johnson, *"art, n.s." Dictionary of the English Language* (1773; accessed July 23, 2024; https://johnsondictionaryonline.com/1773/art_ns).

14. See Neil McKendrick, John Brewer, and J. H. Plumb's influential *The Birth of a Consumer Society: The Commercialization of Eighteenth-Century England* (London: Europa, 1982). J. R. Kellett attributes the waning of the powerful craft guilds in London, over the course of the century, to the influx of craft products from the areas outside of London and the influx of craftsmen from abroad ("The Breakdown of Gild and Corporation Control over the Handicraft and Retail Trade in London," *Economic History Review* 10.3 [1958]: 381–94).

15. Rajani Sudan, *The Alchemy of Empire: Abject Materials and the Technology of Colonialism* (New York: Fordham University Press, 2016).

16. See Joel Mokyr, *The Gifts of Athena: Historical Origins of the Knowledge Economy* (Princeton, NJ: Princeton University Press, 2004).

17. William Eamon also calls these early modern histories of the practical arts "how-to-do-it books" and "self-help books" (*Science and the Secrets of Nature: Books of Secrets in Medieval and Early Modern Culture* [Princeton, NJ: Princeton University Press, 1996] see especially chap. 7).

18. Mokyr calls the period the "Industrial Enlightenment," situated between the seventeenth-century scientific revolution and the nineteenth-century industrial one (*Gifts of Athena* chap. 2).

19. John Locke, *Some Thoughts Concerning Education* (ed. John W. and Jean S. Yolton [Oxford: Clarendon, 1989] 257).

20. See Frans De Bruyn's account of Virgil's poem treated as a scientific treatise in "Reading Virgil's *Georgics* as a Scientific Text: The Eighteenth-Century Debate between Jethro Tull and Stephen Switzer," *ELH* 71.3 (2004): 661–89.

21. Ephraim Chambers, *Cyclopaedia; or, an Universal Dictionary of Arts and Sciences* (London, 1728) ii.

22. See Ronald Paulson's account of Hogarth's entanglement with debates over a British art academy (*Hogarth*, 3 vols. [New Brunswick, NJ: Rutgers University Press, 1991–93] 3:11–16).

23. Larry Shiner, *The Invention of Art: A Cultural History* (Chicago: University of Chicago Press, 2001) 79–80.

24. I am indebted to the histories of craft knowledge that bring histories of science and technology to bear on the history of art, especially Pamela H. Smith, *The Body of the Artisan: Art and Experience in the Scientific Revolution* (Chicago: University of Chicago Press, 2004); Fox, *Arts of Industry;* Matthew Hunter, *Wicked Intelligence: Visual Art and the Science of Experiment in Restoration London* (Chicago: University of Chicago Press, 2013).

25. Most work on the topic of the artisan and the philosopher focuses on early modern Europe. Critical works that anchor the current discussion are Smith, *Body of the Artisan*; Pamela O. Long, *Artisan/Practitioners and the Rise of the New Sciences, 1400–1600* (Corvallis: Oregon State University Press, 2011); and Pamela O. Long, *Openness, Secrecy, Authorship: Technical Arts and the Culture of Knowledge from Antiquity to the Renaissance* (Portland: Oregon State Press, 2011). The lengthy history of discussions over the artisan's role in the scientific revolution may be charted through Paulo Rossi, *Philosophy, Technology, and the Arts* (New York: Harper & Row, 1970); Arthur Clegg, "Craftsmen and the Origin of Science," *Science and Society* 43.2 (1979): 186–201; A. C. Crombie, "Science and the Arts in the Renaissance: The Search for Truth and Certainty, Old and New," *History of Science* 18.42 (1980): 233–46; and James A. Bennett, "The Mechanics' Philosophy and the Mechanical Philosophy," *History of Science* 24.1 (1986): 1–28.

26. Aristotle, *Nicomachean Ethics* VI. See Aristotle's examples of the "wise sculptor" and the "wise maker of statues" (*Nicomachean Ethics* [ed. Roger Crisp [Cambridge: Cambridge University Press, 2000] 108). Aristotle's distinctions are more complex than this: theoretical knowledge contains "episteme" and "nous," and "practical wisdom" and "expertise" are separate forms of knowledge for Aristotle, the first involving praxis or action and the second making or "poiesis." For full account, and for a discussion of the examples I give here, see David Bostock, *Aristotle's Ethics* (Oxford: Oxford University Press, 2000) 75–82. Pamela H. Smith gives a succinct summary of the problem in terms of craft knowledge, *Body of the Artisan* 17. Studies of Aristotelian ethics are significantly more complicated than the binary, part of whose long life I study here, would suggest. See, for example, Tom Angier, *Techne in Aristotle's Ethics: Crafting the Moral Life* (London: Bloomsbury, 2012).

27. Smith, *Body of the Artisan* 7. Richard Sennett explains, "This view, in which the educated generalist dominates the craftsman specialist, reflected a clear hierarchical structure in the Roman state" (*The Craftsman* [New Haven, CT: Yale University Press, 2008] 133). Rossi observes, further, "Thus the opposition between slaves and freemen tended to break down on the basis of technique and science" (*Philosophy* 13).

28. In the current U.S. context, we see this hierarchy everywhere: in the disparaging characterizations of and lesser pay for "manual labor" or "service" jobs; in the separation of "vocational" from other forms of education; in the career paths that lead one away from the hands-on work in which one was trained and into supposedly superior, big-picture jobs of administration or management. We see, too, attempts to flip the hierarchy, arguments for the validity of manual labor and the worthlessness of the intellectual "elites."

29. Bacon, *Advancement* 27–28.

30. Bacon, *Advancement* 94. See Paola Bertucci's opening account of Vulcan and his centrality to the Société des Arts in Paris (*Artisanal Enlightenment: Science and the Mechanical Arts in Old Regime France* [New Haven, CT: Yale University Press, 2017] 2–4).

31. See Antonio Pérez-Ramos's account: "In Bacon's case, his *factum (opus)* is not the artefact which man constructs to mirror Nature and which is therefore designed to lead to propositional knowledge. Rather, Bacon's *factum* is the technical know-how, theoretically more or less informed, which leads to the voluntary (re)production of Nature's phenomena

by the knower/agent" (*Francis Bacon's Idea of Science and the Maker's Knowledge Tradition* [Oxford: Oxford University Press, 1989] 156).

32. Francis Bacon, *The New Organon* (ed. Lisa Jardine and Michael Silverthorne [Cambridge: Cambridge University Press, 2000] 7).

33. Ibid. 227.

34. Ibid. 227–28

35. Fox, *Arts of Industry* 29.

36. Ibid. 35.

37. John Morgan, "Sprat, Thomas (bap. 1635, dd. 1713), bishop of Rochester," *Oxford Dictionary of National Biography* (Oxford: Oxford University Press, 2004; accessed August 1, 2024) www.oxforddnb.com/view/10.1093/ref:odnb/9780198614128.001.0001/odnb-9780 198614128-e-26173.

38. Smith, *Body of the Artisan* 20.

39. The term "maker's knowledge" is from Pérez-Ramos; Bacon, *New Organon* 81.

40. Picciotto, *Labors of Innocence* 1. Picciotto uses the term "corporate body" in reference to the Royal Society, 186. In some ways my account of Royal Society knowledge is closer to William T. Lynch's in *Solomon's Child: Method in the Early Royal Society of London* (Stanford, CA: Stanford University Press, 2001). Lynch considers a Baconian "constructivist" strain of knowledge within the society, focusing on "manual objects" and "a constructivist objectivity" (23). Yet such "constructivism" cannot be separated from problems of embodiment, as Sprat's *History* suggests.

41. Ibid. 10.

42. Ibid. 2.

43. Michael Hunter, *The Royal Society and Its Fellows, 1660–1700: The Morphology of an Early Scientific Institution* (Chalfont St. Giles: British Society for the History of Science, 1982) 25.

44. Rob Iliffe, "Material Doubts: Hooke, Artisan Culture and the Exchange of Information in 1670s London," *British Journal for the History of Science* 28.3 (1995): 285.

45. See especially Iliffe's account of Hooke's work after the Great Fire (ibid. 292–93).

46. Ibid. 289.

47. Fox, *Arts of Industry* 39.

48. Ibid. 39.

49. Steven Shapin, *A Social History of Truth: Civility and Science in Seventeenth-Century England* (Chicago: University of Chicago Press, 1994) 376, 389.

50. Stephen Pumfrey, "Who Did the Work? Experimental Philosophers and Public Demonstrators in Augustan England," *British Journal for the History of Science* 28.2 (1995): 132.

51. Shapin and Schaffer, *Leviathan and the Air-Pump* 65–69; Chico, *Experimental Imagination* 44.

52. Al Coppola, *The Theater of Experiment: Staging Natural Philosophy in Eighteenth-Century Britain* (New York: Oxford University Press, 2016) 10–11.

53. Richard Price, *Labour in British Society: An Interpretive History* (London: Croom Helm, 1986) 15.

54. Ibid.
55. Ibid.
56. Mokyr, *Gifts of Athena* chap. 2.
57. Eamon, *Science and the Secrets of Nature*.
58. Long, *Artisan/Practitioners* 37; Smith, *Body of the Artisan* 59–94.
59. Fox, *Arts of Industry* 16.
60. Ibid.
61. Joseph Moxon, *Mechanick Exercises, or the Doctrine of Handy-Works* (London, 1683–85) preface n.p.
62. On *Mechanick Exercises* (including the plate I describe here), see Fox 31–34.
63. See Margaret Cohen, *The Novel and the Sea* (Princeton, NJ: Princeton University Press, 2010); Jonathan Kramnick, *Paper Minds: Literature and the Ecology of Consciousness* (Chicago: University of Chicago Press, 2018); and Sean Silver, *The Mind Is a Collection: Case Studies in Eighteenth-Century Thought* (Philadelphia: University of Pennsylvania Press, 2015). Cohen's account begins with Odysseus and she focuses on the "practical resourcefulness of the mariner," a key figure in accounts of Enlightenment practical knowledge. See her account of the way such practical knowledge makes its way into form in *Robinson Crusoe*. I turn to her account of subjectivity in Chapter 6.
64. *Craft* is a term of analysis for Cohen, but in considering the maritime context, she skirts the early modern history of craft (as well as its eighteenth-century trajectory) and pulls the term from the writing of Joseph Conrad (*Novel and the Sea* 4). Cohen's primary term is "practical reason," which is at times close to the embodied knowledge I treat here, though crucially "craft" does not have to intersect with "reason" in this book.
65. Edwin Hutchins illuminates the complex genesis of "cognitive ecology," in "Cognitive Ecology," *Topics in Cognitive Science* 2.4 (2010): 705–15; Silver, *The Mind* viii, ix, xi, 5, and 15–18; Kramnick, *Paper Minds* 9–11. When Silver uses "metaphor" to link the conceptual and the haptic, that structure is similarly transhistorical, a formal given (see *The Mind* 11–13).
66. Kramnick *Paper Minds* 3; Silver, *The Mind* viii. Kramnick takes "entanglements" from N. Katherine Hayles (7) but the term is widely used to describe ways in which mind and world are interconnected, sometimes with an origin of the term in the "quantum entanglement" of quantum mechanics. For an example of Silver's use of the term, see viii.
67. The awkwardness is visible both in the incomplete ways in which eighteenth-century authors measure up to a phenomenological model (with which, after all, they are not working) and in critics' need to leave certain aspects of the craft model to the wayside, even when they reference it. See, for example, Abigail Zitin's refusal to acknowledge the endpoint of craft in the object in her discussion of Hogarth (*Practical Form: Abstraction, Technique, and Beauty in Eighteenth-Century* Aesthetics [New Haven, CT: Yale University Press, 2020] chap. 4).
68. The term *mythical norm* is Audre Lorde's from "Age, Race, Class, and Sex: Women Redefining Difference," in *Sister Outsider: Essays and Speeches* (Berkeley, CA: Feminist Press, 1984) 114–23.
69. The relationship between handicraft and Marx's more general thinking about labor has been a matter of some dispute. See, for instance, Sean Sayers, "The Concept of

Labor: Marx and His Critics," *Science and Society* 71.4 (2007): 431–54; and Kaan Kangal, "Sean Sayers' Concept of Immaterial Labor and the Information Economy," *Science and Society* 81.1 (2017): 124–32.

70. In Heidegger's work see especially "The Question Concerning Technology" (*The Question Concerning Technology*, trans. William Lovitt [New York: Harper, 1982]) and "On the Origin of the Work of Art" (*Poetry, Language, Thought*, trans. Albert Hofstadter [New York. Harper & Row, 1971]). Craft plays an influential role in accounts of the Frankfurt School, in which it suggests a stance against a reason-driven totalitarian Enlightenment, most famously in Max Horkheimer and Theodor W. Adorno, *Dialectic of Enlightenment* (Stanford, CA: Stanford University Press, 2002). Benjamin goes further in imagining that craft can redeem technology; see, for instance, "The Storyteller" and "The Work of Art in the Age of Mechanical Reproduction" (*Illuminations*, trans. Harry Zohn [New York: Random House, 1969], 83–109, 217–52).

71. Sennett, *The Craftsman* 11.

72. In the most striking examples, "ecology" can lend a formal relation a positive ethical charge, no matter the subject of the representation. We see this Kramnick's "ecophenomenological" accounts, which usually require that older readings that acknowledge power relations between human entities be ignored (*Paper Minds* 3). Thus, Crusoe can have an ethically-charged relationship to his surroundings, regardless of the colonial overtones or the fact that one aspect of that environment is his servant, Friday. For the dismissal of John Barrell's reading of power relationships in landscape poetry, see Kramnick, *Paper Minds* 63–64; for the description of Swift's Celia as a neutral artisan figure, see Kramnick, *Paper Minds* 86–87 and my reading below, Chapter 2.

73. See Tita Chico's account of the "immodest witness" and Al Coppola's account of a related, bodily observer whose "identity confusion" could result from experiment's "spectacle" (Chico, *Experimental Imagination* 34; Coppola, *Theater of Experiment* 35).

74. Recent literary historians have privileged other origins over the scientific one, as in Mary Poovey's turn to economic writers and Margaret Cohen's to seamen. That said, Sprat's attachment of this way of writing to the new science gave it additional social and epistemological punch, and my interest is in how this scientific association offers writers a starting point for considering other ways in which "experience" might enter language. Mary Poovey, *A History of the Modern Fact* (Chicago: University of Chicago Press, 1998); Cohen, *Novel and the Sea*. For an excellent treatment of "style" as realized in Royal Society publications, see Peter Dear, "*Totius in verba*: Rhetoric and Authority in the Early Royal Society," *Isis* 76.2 (1985): 144–61.

75. Abraham Cowley, "To the Royal Society," in Thomas Sprat, *The History of the Royal Society of London, for the Improving of Natural Knowledge* (London, 1667) n.p.

76. Lynn Festa, *Fiction Without Humanity: Person, Animal, Thing in Early Enlightenment Literature and Culture* (Philadelphia: University of Pennsylvania Press, 2019) 48.

77. Moxon, *Mechanick Exercises* n.p. Fox considers Moxon's text to be the closest anyone came to fulfilling the project of Bacon's projected histories of trades (*Arts of Industry* 39).

78. Peter Dormer, *Art of the Maker: Skill and Its Meaning in Art, Craft, and Design* (London: Thames and Hudson, 1994) 11. David Pye, *The Nature and Art of Workmanship*

(London: Cambridge University Press, 1968) 1. Also see Howard Risatti, *A Theory of Craft: Function and Aesthetic Expression* (Chapel Hill: University of North Carolina Press, 2007).

79. Svetlana Alpers, *The Art of Describing: Dutch Art in the Seventeenth Century* (Chicago: University of Chicago Press, 1983). Although Shapin and Schaffer use Alpers's argument in this way, her fuller account of Dutch painting is much closer to my own argument about representation and craft. See, for example, her account of the Dutch attention to "the crafted surface of their representations" (114). Alpers sets out what is at stake in "handicraft philosophy" when she responds to Foucault's reading of *Las Meninas* in chapter 2. That argument is brought together usefully in her article "Interpretation Without Representation; or, The Viewing of *Las Meninas*," *Representations* 1(1983): 30–42. Alpers identifies Velásquez's painting with "two conflicting modes of representation, each of which constitutes the relation between the viewer and the picturing of the world differently" ("Interpretation" 36). The painting is a combination of Albertian and Northern models; the latter brings us back to the surface of the painting and attends to how the picture is being made. It thus brings in the craftsman, for Velásquez "does not distinguish the liberal art from its craft" ("Interpretation" 33). Hogarth attends to these terms, perhaps due to his connections to Dutch art, as I demonstrate in Chapter 3.

80. Maximillian E. Novak, *Realism, Myth, and History in Defoe's Fiction* (Lincoln: University of Nebraska Press, 1983) 1, 31. Mary Baine Campbell, *Wonder and Science: Imagining Worlds in Early Modern Europe* (Ithaca, NY: Cornell University Press, 1999) 201.

81. Shapin and Schaffer, *Leviathan* 60–63.

82. See Helen Thompson's work with realism and representation in *Fictional Matter: Empiricism, Corpuscles, and the Novel* (Philadelphia: University of Pennsylvania Press, 2017) 11–13.

83. Ilse Vickers, *Defoe and the New Sciences* (Cambridge: Cambridge University Press, 1996).

84. Daniel Defoe, *Robinson Crusoe* (1719; New York: W. W. Norton, 1994) 89.

85. Dear, *"Totius in verba"* 152.

86. Defoe, *Robinson Crusoe* 89.

87. Ibid. 90.

88. Ibid.

89. Montesquieu, *Spirit* 1.

90. David Hume, *An Abstract of a Book Lately Published; Entitled A Treatise of Human Nature* in *David Hume: A Treatise of Human Nature*, ed. David Fate Norton and Mary J. Norton, 2 vols. (Oxford: Oxford University Press, 2007) 407.

91. See, for example, Manicas, *History and Philosophy*; and Fox, Porter, and Wokler, eds., *Inventing Human Science* esp. the introduction.

92. William Dampier, *Memoirs of a Buccaneer: Dampier's New Voyage Round the World, 1697* (Mineola, NY: Dover, 1968).

93. See especially Jason H. Pearl, "Geography and Authority in the Royal Society's Instructions for Travelers," in *Travel Narratives, the New Science, and Literary Discourse, 1569–1750*, ed. Judy A. Hayden (Burlington, VT: Ashgate, 2012) 71–86.

94. James Douglas, "*Hints* offered to the consideration of Captain Cooke, Mr Bankes, Doctor Solander, and the other Gentlemen who go upon the Expedition on Board the

Endeavour," in *The Journals of Captain James Cook*, vol. 1, *The Voyage of the* Endeavour, *1768–1771*, ed. J. C. Beaglehole (Cambridge: Hakluyt Society and Cambridge University Press, 1955) 514–19.

95. David Carrithers, "The Enlightenment Science of Society," in *Inventing Human Science*, ed. Fox, Porter, and Wokler, 239.

96. Margaret T. Hodgen, *Early Anthropology in the Sixteenth and Seventeenth Centuries* (Philadelphia: University of Pennsylvania Press, 1964) 189–90.

97. Gregory Bateson, *Steps to an Ecology of Mind* (New York: Ballantine, 1972) 113.

98. Clifford Geertz, *Local Knowledge: Further Essays in Interpretive Anthropology* (New York: Basic, 1985).

99. Michael Jackson, introduction to *Things as They Are: New Directions in Phenomenological Anthropology*, ed. Michael Jackson (Bloomington: Indiana University Press, 1996) 18.

100. Pierre Bourdieu, *Outline of a Theory of Practice*, trans. Richard Howard (New York: Cambridge University Press, 1977). For the explicit turn toward the body, see Bourdieu's source in Marcel Mauss, "Techniques of the Body," *Economy and Society* 2.1 (1973): 70–88. This is one element of Bourdieu's response to structuralism: people do not operate by hidden social rules but instead in terms of an interaction between actor and structure. Pierre Bourdieu, Jean-Claude Chamboredon, and Jean-Claude Passeron, *The Craft of Sociology: Epistemological Preliminaries*, ed. Beate Krais, trans. Richard Nice (New York: Walter de Gruyter, 1991).

101. Certeau, *The Practice of Everyday Life*, trans. Steven Rendall (Berkeley: University of California Press, 1984), xv. Later in the book de Certeau turns to what he calls a "Baconian" "distiction" in Diderot's account of "art" versus "science" (66).

Chapter 1

1. Ian Watt, *The Rise of the Novel: Studies in Defoe, Richardson, and Fielding* (London: Penguin, 1963) 108.

2. Daniel Defoe, *Robinson Crusoe* (1719; New York: W. W. Norton, 1994) 38.

3. Ibid. 117, 73, 167.

4. For definitions of Defoe's realism through the objects in his prose, see, for example, Wolfram Schmidgen on items from the ship; Cynthia Sundberg Wall on pots; and Jenny Davidson on shoes. Schmidgen, *Eighteenth-Century Fiction and the Law of Property* (Cambridge: Cambridge University Press, 2002) 49–50; Wall, *The Prose of Things: Transformations of Description in the Eighteenth Century* (Chicago: University of Chicago Press, 2006); Davidson, "The 'Minute Particular' in Life-Writing and the Novel," *Eighteenth-Century Studies* 48.3 (2015): 275.

5. See Lincoln Faller's account in *Crime and Defoe: A New Kind of Writing* (Cambridge: Cambridge University Press, 1993). Faller registers a version of this divide when he comments that there are fewer solid particulars in the criminal novels than in *Robinson Crusoe* and that this is an odd absence given their attention to "step by step" description (*Crime* 31, 39). Faller opts to read the lack of "formal realism" as an invitation to consider form on a larger scale, by way of criminal biography and Defoe's deviations from its social resolution for punishment.

6. Ilse Vickers, *Defoe and the New Sciences* (Cambridge: Cambridge University Press, 1996) 2. For her account of Morton, see Chapter 3.

7. Michael Shinagel, *Daniel Defoe and Middle-Class Gentility* (Cambridge, MA: Harvard University Press, 1968) 20.

8. Shinagel notes that Defoe's entrance into Morton's academy came in lieu of an apprenticeship in his father's trade (*Daniel Defoe* 18).

9. Daniel Defoe, *The Complete English Tradesman in Familiar Letters, Directing Him in All the Several Parts and Progressions of Trade* (*Religious and Didactic Writings of Daniel Defoe*, vol. 7, ed. W. R. Owens and P. N. Furbank [London: Pickering & Chatto, 2007]). Hereafter cited in the text as *CET*.

10. Defoe's biographers John Richetti and Paula R. Backscheider adopt the term, though both observe that the work contains various forms of knowledge. Richetti, *The Life of Daniel Defoe* (Malden, MA: Blackwell, 2005) 155; and Backscheider, *Daniel Defoe: Ambition and Invention* (Louisville: University Press of Kentucky, 1986) 65.

11. Daniel Defoe, *A Review of the State of the English Nation Defoe's Review: Facsimile Book 6 of Vol. III (Jan. 1, 1706–May 4, 1706)* (ed. Arthur Wellesley Secord [New York: Columbia University Press, 1938] 6–7).

12. Richard A. Barney, *Plots of Enlightenment: Education and the Novel in Eighteenth-Century England* (Stanford, CA: Stanford University Press, 1999) 54.

13. A number of prominent educational tracts (less famous than Peacham's) bore similar names, e.g., Richard Brathwaite, *The English Gentleman* (London, 1630), and Jean Gailhard, *The Compleat Gentleman* (London, 1678).

14. We should not overestimate the novelty of Locke's claims, as Barney has argued. Locke's "treatise consolidated many of the pedagogical innovations proposed since the mid-seventeenth century by authors such as Comenius, Bathsua Makin, and Jean Gailhard" (*Plots of Enlightenment* 38).

15. John Locke, *Essay Concerning Human Understanding* (ed. Peter H. Nidditch [Oxford: Oxford University Press, 1979]) 107, 108.

16. John W. and Jean S. Yolton set aside Locke's attention to the body as a peculiarity of his medical training in their introduction to *Some Thoughts Concerning Education* (ed. John W. and Jean S. Yolton [Oxford: Clarendon, 1989]) 35. Both Barney and Jenny Davidson read the attention to the body as a first step only, a condition for later developing the mind (Barney, *Plots of Enlightenment* 39 ff.; Davidson, *Breeding: A Partial History of the Eighteenth Century* [New York: Columbia University Press, 2009] chap. 2).

17. John Locke, *Some Thoughts Concerning Education* (ed. John W. and Jean S. Yolton [Oxford: Clarendon, 1975]) 125, 214. Hereafter cited in the text as *E*.

18. Thomas Sprat, *The History of the Royal-Society of London, For the Improving of Natural Knowledge* (London, 1667) 85.

19. Defoe compares the tradesman to the navigator in the *Tradesman* (232, 245–46). For a history of navigation as craft, see Margaret Cohen, *The Novel and the Sea* (Princeton, NJ: Princeton University Press, 2010) chap. 1.

20. The full sentence reads: "Quod medicum est medici, *Tractent fabrilia fabri*" (Horace, *Epistles* II.i.115–16; *Satires. Epistles. The Art of Poetry*, ed. H. Rushton Fairclough [Cambridge, MA: Harvard University Press, 1926]).

21. Horace, *Satires* II.i.116–17.

22. Sprat claims the "Tradesman" among his sources for the plain style (*History* 67).

23. Ibid. 112, 113.

24. On the limitations of Defoe's view, see Janet Sorensen, *Strange Vernaculars: How Eighteenth-Century Slang, Cant, Provincial Languages, and Nautical Jargon Became English* (Princeton, NJ: Princeton University Press, 2017) esp. chap. 8.

25. Mary Poovey argues that the plain style originates not in philosophical but in economic writings, like those of the seventeenth-century writer, Thomas Mun (*A History of the Modern Fact* [Chicago: University of Chicago Press, 1998] 66–68). On other sources of plain language, see Poovey, *Modern Fact* 81–82.

26. Ibid. 102.

27. Ibid. 100.

28. Daniel Defoe, *A Supplement to the Complete English Tradesman* (*Religious and Didactic Writings of Daniel Defoe*, vol. 7, ed. W. R. Owens and P. N. Furbank [London: Pickering & Chatto, 2007]) 275.

29. Sandra Sherman, "Commercial Paper, Commercial Fiction: 'The Complete English Tradesman' and Defoe's Reluctant Novels," *Criticism* 37.3 (1995): 405–6.

30. Ibid. 406.

31. Since Samuel Monk's reading of the novel established its relation to the bildungsroman, critics repeatedly have returned to the term (introduction to *Colonel Jack*, by Daniel Defoe, ed. Samuel Holt Monk [Oxford: Oxford University Press, 1970]). John J. Richetti rightly notes that by contrast with the nineteenth-century hero, Jack does not develop (*Defoe's Narratives: Situations and Structures* [Oxford: Clarendon, 1975] 151). Erin Mackie disagrees, reviving the term as a description of Jack's trajectory toward his identity as a gentleman (*Rakes, Highwaymen, and Pirates: The Making of the Modern Gentleman in the Eighteenth Century* [Baltimore: Johns Hopkins University Press, 2009] 17–21). Richard A. Barney's account of the "narrativization of educational writing" is closer to what I consider here. Barney describes (via Pierre Bourdieu and Judith Butler) an "improvisational" subjectivity visible in educational theory and in the early novel. The link between the two genres of writing for him depends upon a similar subject, consistently negotiating the terms for applying an "internalization system of social codes" to their particular circumstances (*Plots of Enlightenment* 15–16). Barney considers *Robinson Crusoe*, but *Colonel Jack* offers an even clearer case of a Lockean educational paradigm.

32. Defoe's account becomes complicated along just the lines Davidson suggests in *Breeding*, where eighteenth-century ideas of "custom" and education seem both detached from birth and blood and, at the same time, in service to them (see esp. chap. 1).

33. David Blewett does acknowledge that this thesis is not proven through Jack himself but through the Captain. He also acknowledges that this does not seem to be Defoe's point in *Moll Flanders*. Two of Defoe's most famous criminals, then, call for another explanation of the role of education (*Defoe's Art of Fiction: Robinson Crusoe, Moll Flanders, Colonel Jack, and Roxana* [Toronto: University of Toronto Press, 1979] 104, 62).

34. Daniel Defoe, *The History and Remarkable Life of the Truly Honourable Col. Jacque, Commonly Called Col. Jack* (ed. Gabriel Cervantes and Geoffrey Sill [Peterborough: Broadview, 2016] 68). Hereafter cited in the text as *CJ*.

35. See Frank McLynn, *Crime and Punishment in Eighteenth-Century England* (London: Routledge, 1989) 7.

36. Roland Barthes, "The Reality Effect" (1969; in *The Rustle of Language*, trans. Richard Howard [Berkeley: University of California Press, 1989] 141–48).

37. Wall reminds us that the "traditionally defined focus of description obviously *is* objects of one sort or another" (*Prose* 2). Also see Wall's acknowledged indebtedness to Bill Brown's thinking on "things" (*Prose* xi). Wolfram Schmidgen observes that definitions of description that take Marxism or structuralism as their source may end up in much the same position, opposing "description" to "narrative" and assuming "that description is inherently tied to the referential illusion and arrives as a significant literary mode with the large-scale reification of social relations ushered in by the industrial revolution" (*Eighteenth-Century Fiction* 60). He argues that Defoe's realism is better understood by considering that object and narrative could mingle in the world of mercantilism.

38. Novak, *Realism, Myth, and History in Defoe's Fiction* (Lincoln: University of Nebraska Press, 1993) 2. Novak links this to later novelistic and painterly practice: "Like later realists, Defoe, too, sought a language that would be direct and completely descriptive of the perceived object" (8–9).

39. Mary Baine Campbell, *Wonder and Science: Imagining Worlds in Early Modern Europe* (Ithaca, NY: Cornell University Press, 1999) 201; Tita Chico, "Minute Particulars: Microscopy and Eighteenth-Century Narrative," *Mosaic* 39.2 (2006): 143–61.

40. Dorothy Van Ghent, *The English Novel: Form and Function* (1952; New York: Harper & Row, 1961), qtd. in Wall, *Prose* 1.

41. Wall, *Prose* chap. 3.

42. See Abraham Cowley's prefatory poem to Sprat's *History of the Royal Society*: "To the Royal Society," *The History of the Royal-Society of London, for the Improving of Natural Knowledge* (London, 1667) n.p.

43. Svetlana Alpers, *The Art of Describing: Dutch Art in the Seventeenth Century* (Chicago: University of Chicago Press, 1983) 79–80.

44. Two important exceptions are the "performability" or "function" that Margaret Cohen and Lynn Festa (respectively) oppose to realism based on visual detail. Defoe's ultimate experimentation with plain language diverges from these accounts, in that he pulls apart and resolves separately the relation between observer and observed, an important piece of moving the natural philosophical model into the social world. See Cohen, *Novel and the Sea* 72; Festa, *Fiction Without Humanity: Person, Animal, Thing in Early Enlightenment Literature and Culture* (Philadelphia: University of Pennsylvania Press, 2019) 113.

45. Peter Dear, "*Totius in verba*: Rhetoric and Authority in the Early Royal Society," *Isis* 76.2 (1985): 145–61. This account is a significantly more precise than the better-known one (at least in literary circles) in Steven Shapin and Simon Schaffer, *Leviathan and the Air-Pump: Hobbes, Boyle, and the Experimental Life* (Princeton, NJ: Princeton University Press, 1985). There, Shapin and Schaffer assume a straightforward analogy between the "'somewhat prolix'" accounts of Boyle's experiments and painting, citing Alpers

(*Leviathan* 62). "Evidently, the Dutch were trying to achieve by way of picturing what the English were attempting through the reform of prose" (*Leviathan* 62, n83). See Helen Thompson's discussion and critique of Shapin and Schaffer on mechanism and its implications for their account of representation in *Fictional Matter: Empiricism, Corpuscles, and the Novel* (Philadelphia: University of Pennsylvania Press, 2017) 11–13.

46. Dear, *"Totius in verba"* 153.
47. Ibid.
48. Ibid.
49. Ibid.
50. Ibid. 152.
51. Stephen Shapin, "Pump and Circumstance: Robert Boyle's Literary Technology," *Social Studies of Science* 14 (1984): 481–519, qtd. in Dear *"Totius in verba"* 152, n20.
52. A third example comes from "a Court, which goes out of *Grace-Church-street* into *Lombard-street*, where the *Quaker's-Meeting House* is." It begins this way, and again the language is all "force" and "Thrust": "Keep up, says *Will* to me, be nimble, and as soon as he had said so, he flyes at the young Man, and Gives him such a violent Thrust, that push'd him forward with too great a force for him to stand, and as he strove to recover, the Threshold took his Feet, and he fell forward into the other part of the Court, as if he had flown in the Air, with his Head lying towards the *Quaker's-Meeting-House*" (*CJ* 110).
53. Thompson, *Fictional Matter* 20.
54. Blewett reads Jack's entire character and his participation in business and trade as ironic (*Defoe's Art* chap. 4). My analysis is more in line with that of critics like Richetti and Faller, who have pushed back against the account of simple irony and who examine the way in which Defoe poses questions without offering firm answers and lays out logics without necessarily standing behind them or wrapping them up neatly. As Richetti puts it, "Defoe's narratives may be said to exploit ideas much more than they resolve them or even articulate them fairly" (*Defoe's Narratives* 145). Faller offers a related idea of Defoe's complexity: "All his novels are meditations on the incompleteness of truth, or, rather, on the incompleteness of our ability to recognize and grasp what truths we encounter" (*Crime and Defoe* 253).
55. Tita Chico, *The Experimental Imagination: Literary Knowledge and Science in the British Enlightenment* (Stanford, CA: Stanford University Press, 2018) 47; Al Coppola, *The Theater of Experiment: Staging Natural Philosophy in Eighteenth-Century Britain* (New York: Oxford University Press, 2016) 35-36.
56. Amit Yahav-Brown writes of "social organization" in *Moll Flanders* and Carl Fisher uses the term "societal character" to specify Defoe's interest in the South Sea Bubble ("At Home in England, or Projecting Liberal Citizenship in *Moll Flanders*," *Novel: A Forum on Fiction* 35.1 [2001] 25); "'The Rage of the Street': Crowd and Public in Defoe's *Moll Flanders*," *Historical Boundaries, Narrative Forms: Essays on British Literature in the Long Eighteenth Century in Honor of Everett Zimmerman*, ed. Lorna Clymer and Robert Mayer (Newark, DE: University of Delaware Press, 2007) 74.
57. Faller, *Crime and Defoe* 5.
58. Ibid. 71.
59. See Davidson's discussion (*Breeding* 42–48).

60. The punishment is never named, but it is the "mobile punishment" that McLynn calls "the most common form of public flagellation . . . 'at the cat's tail'" (*Crime and Punishment* 281).

61. See J. M. Beattie's account of the reconsideration of corporal punishment and reassessment of crime that is visible through court proceedings, if not yet in public debate (*Crime and the Courts in England, 1660–1800* [Oxford: Clarendon, 1986] lx).

62. Defoe writes within the larger reform movement that Paul B. Shoemaker records (*Prosecution and Punishment: Petty Crime and the Law in London and Rural Middlesex, c. 1660–1725* [Cambridge: Cambridge University Press, 1991] 162). On the Bloody Code, see McLynn, *Crime and Punishment* xi. The anonymous author of *Hanging Not Punishment Enough* writes in just this spirit, when he supports even bloodier punishments than the Bloody Code lays out [(London, 1791) 1).

63. Richetti, *Defoe's Narratives* 152.

64. The slave plantation reflects debates over punishment in England. On reform that took punishment beyond terror, see Shoemaker, *Prosecution and Punishment* 193.

65. In his reading of this episode, George Boulukos cites an earlier scene of gratitude and enslavement as a precedent for what Jack does here. I'm suggesting that the source is much deeper in Jack's body and mind, occurring well before slavery enters the novel (*The Grateful Slave: The Emergence of Race in Eighteenth-Century British and American Culture* [Cambridge: Cambridge University Press, 2008] 91).

66. Richetti, *Defoe's Narratives* 149.

67. Faller observes that "whippings, beatings, and fights" are a "reoccur[ing] motif" (*Crime and Defoe* 172).

Chapter 2

1. Frans De Bruyn, "Reading Virgil's *Georgics* as a Scientific Text: The Eighteenth-Century Debate Between Jethro Tull and Stephen Switzer," *ELH* 71.3 (2004): 661–89. In such arguments writers like Jethro Tull and Stephen Switzer not only quibble over Virgil's particular advice on planting but over concepts like "experience" and how both poetry and science should admit of and relate it (662). For other connections between the georgic and the new science, see David Alff, *The Wreckage of Intentions: Projects in British Culture, 1660–1730* (Philadelphia: University of Pennsylvania Press, 2017) esp. ch. 2; Juan Christian Pellicer, "The Georgic at Mid-Eighteenth Century and the Case of Dodsley's 'Agriculture,'" *Review of English Studies* 54.213 (2003): 67–93.

2. Francis Bacon, *The Advancement of Learning* (1605; ed. Stephen Jay Gould [New York: Random House, 2001]) 158.

3. Ephraim Chambers, *Cyclopaedia; or, an Universal Dictionary of Arts and Sciences* (London, 1728) ii.

4. "Art, *N.* (1), Etymology," *Oxford English Dictionary* (Oxford: Oxford University Press, June 2024) https://doi.org/10.1093/OED/4658965780.

5. John Smallman Gardiner, *The Art and the Pleasures of Hare-Hunting. In Six Letters to a Person of Quality* (London, 1750); E. T., *The Art and Mystery of Vintners and Wine-Coopers. Containing Approved Directions for the Preserving and Curing All Manner and Sorts of Wines* (London, 1734); John Mitchell, Ars Scribendi Sine Penna: *or, the Art of Taking*

Down Sermons, Trials, Speeches, &c. Verbatim, Without Pen and Ink, and Upon One Page (London, 1783).

6. Edward Bysshe, "The Preface," *The Art of English Poetry*, 1st ed. (London, 1702) n.p.; Edward Bysshe, "The Preface," *The Art of English Poetry*, 8th ed. (London, 1737) n.p.; Alexander Pope, *The Dunciad* (London, 1728) III, l. 136. A. Dwight Culler observes that Bysshe at least supervised revisions as late as 1708, though there not evidence that he continued to do so after that date ("Edward Bysshe and the Poet's Handbook," *PMLA* 63.3 [1948]: 861). On Bysshe also see Stephen Jarrod Bernard, "Edward Bysshe and *The Art of English Poetry*: Reading Writing in the Eighteenth Century," *Eighteenth-Century Studies* 46.1 (2012): 113–29.

7. Bysshe, *Art of Poetry* (1737).

8. Culler, "Edward Bysshe" 860; Bysshe, *Art of Poetry* (1737) 2.

9. Bysshe, *Art of Poetry* (1737) n.p.

10. "Art, N. (1)," *Oxford English Dictionary*.

11. We know from Joseph Spence that Duck had access to a small collection of books (including this one) from his friend on the farm, Lavington, who had been to service in London (*A Full and Authentick Account of Stephen Duck, the Wiltshire Poet* [London, 1731] 7–9).

12. See William J. Christmas, *The Lab'ring Muses: Work, Writing, and the Social Order in English Plebian Poetry, 1730–1830* (Newark: University of Delaware Press, 2001) 25.

13. Clifford Siskin, *The Work of Writing: Literature and Social Change in Britain, 1700–1830* (Baltimore: Johns Hopkins University Press, 1998).

14. William J. Christmas, "Introduction: An Eighteenth-Century Laboring-Class Tradition," *Eighteenth Century* 42.3 (2001): 187.

15. William J. Christmas, introduction to *Eighteenth-Century English Labouring-Class Poets, 1700–1800*, ed. William J. Christmas (London: Pickering & Chatto, 2003) 1:xxii–xxiii.

16. For a rare critical account of Spence's biography, see Jennifer Batt, "From the Field to the Coffeehouse: Changing Representations of Stephen Duck," *Criticism* 47.4 (2005): 451–70.

17. Spence, *Full and Authentick* 11.

18. Ibid.

19. Jonathan Swift, "On Stephen Duck, the Thresher, and favourite Poet: A Quibbling Epigram," *Poems on Several Occasions* (Dublin, 1735) 278. For his part, Pope commented on the idea that Duck might be a contender for the Laureate: "For Duck can *thresh*, you know, as well as *write*" ("An Epigram," *Grub-Street Journal* 46 [November 19, 1730] n.p.).

20. Siskin, *Work of Writing* 120, 125. As Clifford Siskin puts it, "Wordsworth returns to the land, but not to till it" (*Work of Writing* 127). Siskin's target is the Romantic poet and the criticism that he sees as its heir: the close reading of New Criticism. This agenda means that he is inclined to note the eighteenth-century georgic's relevance but also skip over engagements with the concept of labor that occur between Dryden and Wordsworth.

21. Joanna Picciotto, *Labors of Innocence in Early Modern England* (Cambridge, MA: Harvard University Press, 2010). Katarina O'Briain reminds us that Dryden's translation is steeped in Royal Society ways of knowing, gathered firsthand through Dryden's own

natural philosophical education ("Dryden's Georgic Fictionality," *Eighteenth-Century Fiction* 30.3 [2018]: 317–38).

22. John Chalker, *The English Georgic: A Study in the Development of a Form* (Baltimore: Johns Hopkins University Press, 1969) 10.

23. Ibid.

24. Stephen Duck, *The Thresher's Labour* (1730; *Eighteenth-Century English Labouring Class Poets, 1700–1800*, ed. William Christmas, vol. 1 [London: Pickering & Chatto, 2003] ll. 13–14). Hereafter cited in the text as *TL* and by line number. There has little agreement on the genre of Duck's poem, and we may conclude up front that it is "mixed," including pastoral (or antipastoral) and passages tied to landscape poetry. I begin by discussing the poem as a georgic, because it makes clear what Duck's poem has to say about labor and natural philosophy, a point obscured by considering it only as a pastoral. The tradition of classifying the poem as pastoral begins with Raymond Williams, *The Country and the City* (Oxford: Oxford University Press 1973), and continues through John Barrell, *The Dark Side of the Landscape: The Rural Poor in English Painting 1730–1840* (Cambridge: Cambridge University Press, 1983); Michael McKeon, "Surveying the Frontier of Culture: Pastoralism in Eighteenth-Century England," *Studies in Eighteenth-Century Culture* 26.1 (1997): 7–28; and James Mulholland, "'To Sing the Toils of Each Revolving Year': Song and Poetic Authority in Stephen Duck's 'The Thresher's Labour,'" *Studies in Eighteenth-Century Culture* 33.1 (2004): 153–74. Critics who have classified the poem instead as a georgic include Steve Van-Hagen, "Literary Technique, the Aestheticization of Laboring Experience, and Generic Experimentation in Stephen Duck's 'The Thresher's Labour,'" *Criticism* 47.4 (2005): 421–50; John Goodridge, *Rural Life in Eighteenth-Century English Poetry* (Cambridge: Cambridge University Press, 1995); Bridget Keegan, "Georgic Transformations and Stephen Duck's 'The Thresher's Labour,'" *SEL: Studies in English Literature, 1500–1900* 41.3 (2001): 545–62; Corey E. Andrews, "'Work' Poems: Assessing the Georgic Mode of Eighteenth-Century Working-Class Poetry," *Experiments in Genre in Eighteenth-Century Literature*, ed. Sandro Jung (Ghent, Belgium: Academia Press, 2011) 105–33.

25. David Fairer, "'Where Fuming Trees Refresh the Thirsty Air': The World of Eco-Georgic," *Studies in Eighteenth-Century Culture* 40.1 (2011): 205. In addition to this list of georgic principles, Fairer lists "*care* in husbandry." Although neither laboring georgic speaks to the farmer, both take up and reconfigure Virgil's "care" or "curas," as in Duck's account of the careful movements of tools.

26. Joseph Addison, "An Essay on the Georgics" (1697; *Poems: The Works of Virgil in English, 1697*, ed. William Frost [Berkeley: University of California Press, 2000] 5:145–54).

27. Kevis Goodman, *Georgic Modernity and British Romanticism: Poetry and the Mediation of History* (New York: Cambridge University Press, 2004) 21.

28. Keegan notes that the very presence of the speaking laborer means that the poem's georgic identity is, in a sense, over before it begins ("Georgic Transformations" 546–47).

29. Goodridge, *Rural Life* 6.

30. Barrell, *Dark Side*.

31. Margaret Anne Doody uses just these terms to describe Duck's innovation: the thresher "has risen from being a figure in a pastoral landscape to the position of speaker

who can tell us what that position feels like" (*The Daring Muse: Augustan Poetry Reconsidered* [Cambridge: Cambridge University Press, 1985] 106). His poem offers a kind of "ideological resistance," as Christmas puts it, to the world of work he experienced (*Lab'ring Muses* 22).

32. Goodman, *Georgic Modernity* 27.
33. Addison, "Essay on the Georgics" 151.
34. Goodman, *Georgic Modernity* 26.
35. Ibid. 28.
36. Sprat, *History of the Royal Society* 21.
37. Addison, "Essay on the Georgics" 146.
38. See, for instance, Gary Lenhart, *The Stamp of Class: Reflections on Poetry and Social Class* (Ann Arbor: University of Michigan Press, 2006) 12; Barrell, *Dark Side*, 7. VanHagen uses the term *raw experience* in describing Duck's poetry ("Literary Technique" 439). Even Goodridge, one of the very best readers of Duck to date, offers the strange assessment that because Duck's threshing precludes poetry, Duck has no choice but to simply write out a description of his working experience. Although Goodridge carefully counters critics who look for "authenticity" in Duck, this would appear to contain a similar kind of assumption about language that cannot quite divorce itself from the merely descriptive, the plain, the near to experience (*Rural Life* 16).
39. Picciotto, *Labors of Innocence* 129–87; Picciotto takes the phrase from Sprat.
40. "Dedication of the Georgics to the Earl of Chesterfield," in *The Works of John Dryden*, vol. 5, *The Works of Virgil in English, 1697*, edited by William Frost (Berkeley: University of California Press, 1988) 137.
41. Genesis 3:11, King James Version; John Milton, *Paradise Lost* (1674; London: Cambridge University Press, 1974) 11.434.
42. Virgil, *Georgics* (*Eclogues, Georgics, Aeneid, Books 1–6*, trans. H. Rushton Fairclough [Cambridge, MA: Harvard University Press, 1994] I, l. 293); Dryden, John, and Virgil, "The First Book of the Georgics," in *The Works of John Dryden*, vol. 5, *Poems, 1697*, ed. William Frost, 5:155–79 (Berkeley: University of California Press, 1987) I. I, 72.
43. See Williams, *Country and the City* chap. 7.
44. Barrell, *Dark Side* 2.
45. Barrell describes Duck as depicting the situation "better than I can" (Barrell, *Dark Side* 118); Williams has Duck's account of the feast step in to correct the abundance of the country house poem (32).
46. Ibid. 5. Christmas calls Duck's strategy "ideology critique." Christmas, *Lab'ring Muses* 24.
47. Goodman, *Georgic Modernity* 40. Goodman draws on an early version of Picciotto's argument here.
48. Andrews, "'Work' Poems" 118.
49. John Barrell, *Poetry, Language, and Politics* (Manchester: Manchester University Press, 1988) 103–4. Also see Goodman on Thomson (*Georgic Modernity* 29, 63) and my discussion of Thomson's "philosophic eye" below.
50. John Barrell, *English Literature in History, 1730–80: An Equal, Wide Survey* (London: Hutchinson, 1983) 33.

51. This would appear to be an extension of an older pastoral tradition with a "land yielding of itself," as in Jonson's Penshurst where "their work is all done for them by a natural order" (Williams, *Country and the City* 32).

52. Barrell, *Poetry, Language, and Politics* 115.

53. Ibid. 105.

54. See Pellicer's analysis of Dodsley's "Agriculture," which suggests that Dodsley arrives at a related position, unwilling to support the gentlemanly vision he begins to offer in the poem ("Georgic at Mid-Eighteenth" 83–87).

55. Goodridge, *Rural Life* 37.

56. Jacques Rancière, "The Aesthetic Dimension: Aesthetics, Politics, Knowledge," *Critical Inquiry* 36.1 (2009): 8. Rancière's example is a joiner who, during the French Revolution of 1848, describes looking out the window from a house where he is laying a floor. As the philosopher puts it, "The perspectival gaze has long been associated with mastery and majesty. But in this case, it is reappropriated as a means of disrupting the adequation of a body and an ethos" (7). Duck offers such a disruption, but he does so by upending the gaze itself.

57. "Easily affected, sensitive," "Tender, *Adj. & N.* (4)." *Oxford English Dictionary* (Oxford: Oxford University Press, December 2023) https://doi.org/10.1093/OED/3831196868.

58. Ralph Cohen, "Notes on the Teaching of Eighteenth-Century Poetry of Natural Description," *Teaching Eighteenth-Century Poetry*, ed. Christopher Fox (New York: AMS Press, 1990) 90. I do not believe that Cohen moves far enough into the poem, however, in evaluating whether it offers a firm ideological split with the pastoral mode.

59. See, for instance, Mulholland, "To Sing the Toils" 157; Goodridge, *Rural Life* 47; Keegan, "Georgic Transformations" 556.

60. See Van-Hagen who remarks on Duck's familiarity with Pope and with the journals containing these debates ("Literary Technique" 421–50). The "realism" debates also frame Barrell's account in *Dark Side* 1–2.

61. [Thomas Tickell], *The Guardian* 22 (April 6, 1713) 144.

62. [Thomas Tickell], *The Guardian* 30 (April 15, 1713) 194.

63. [Thomas Tickell], *The Guardian* 28 (April 13, 1713) 183.

64. [Thomas Tickell], *The Guardian* 23 (April 1, 1713) 150.

65. Keegan, "Georgic Transformations" 545.

66. Van-Hagen, "Literary Technique" 430.

67. Goodman, *Georgic Modernity* 11.

68. Goodridge, *Rural Life* 35.

69. Michael Roberts, "Sickles and Scythes: Women's Work and Men's Work at Harvest Time," *History Workshop* 7 (1979): 9.

70. "Just, Adj." *Oxford English Dictionary* (Oxford: Oxford University Press, March 2024) https://doi.org/10.1093/OED/1171573720.

71. David Fairer emphasizes the georgic's practical knowledge in its tendency to "look beneath efficient systems at the mechanisms that are at work" ("Georgic," *The Oxford Handbook of British Poetry, 1660–1800*, ed. Jack Lynch [Oxford: Oxford University Press, 2016] 471). Duck does something related with the bodily scene of labor here.

72. "Stroke, n.1," *Oxford English Dictionary* (Oxford: Oxford University Press, March 2024) https://doi.org/10.1093/OED/8492985351.

73. Robert Dodsley, "Agriculture," *Public Virtue: A Poem* (London, 1753) l. 100.

74. Christmas, *Lab'ring Muses* 23. Christmas argues that for laboring poets "writing [is] a productive means to replace the drudgery of their previous tasks."

75. E. P. Thompson, "Time, Work-Discipline, and Industrial Capitalism," *Past and Present* 38 (1967): 61.

76. Thompson, "Time" 71.

77. Karl Marx, *Capital* (1867; trans. Ben Fowkes, New York: Vintage, 1977) 1:285, 290, 287.

78. Indeed, Sean Sayers claims that Marx does not actually believe in the model of craft per se ("The Concept of Labor: Marx and His Critics," *Science and Society* 71.4 [2007]: 449–50).

79. Bridget Keegan, "Cobbling Verse: Shoemaker Poets of the Long Eighteenth Century," *Eighteenth Century* 42.3 (2001): 201.

80. Patricia Akhimie, *Shakespeare and the Cultivation of Difference: Race and Conduct in the Early Modern World* (New York: Routledge, 2018) 29.

81. Ibid. 27.

82. Ibid.

83. In raising this anxiety, the figure of the Ethiopian also presses on the expansionist, imperial logic of the georgic poem as Karen O'Brien has discussed it, in its ability to allow readers to link seamlessly the country, the city, and empire. These "aesthetic and moral links" seem troubling in Duck's poem, producing not so much "ideological separation" between country and city (160) as a strange, otherworldly doubling manifested on the body itself (161). O'Brien, "Imperial Georgic, 1660–1789," in *The Country and the City Revisited: England and the Politics of Culture, 1550–1850*, ed. Gerald MacLean, Donna Landry, and Joseph P. Ward (Cambridge: Cambridge University Press, 1999).

84. Mary Collier, *The Woman's Labour: An Epistle to Mr. Stephen Duck* (in *Eighteenth-Century English Labouring-Class Poets*, ed. Christmas, vol. 1) ll. 3–4. Hereafter cited in the text as *WL* and by line number.

85. Hannah Arendt, *The Human Condition* (1958; Chicago: University of Chicago Press, 1998) 153–67. Richard Sennett responds to this position in the introduction to *The Craftsman* (New Haven, CT: Yale University Press, 2008).

86. Goodridge, *Rural Life* 33.

87. Roberts observes that it was precisely this sort of manual work that preserved women's "economic importance" in the period ("Sickles and Scythes" 11).

88. This suggests a different relation between close reading and the work of writing than the one Siskin marks out in his analysis. Far from merely extractive and pleasure driven, Collier's criticism of Duck works to unpack "modern power" (Siskin, *Work of Writing* 129).

89. Here Collier's borrowing performs an experience for the reader that imitates what Audre Lorde calls the "mythical norm," where the distance is felt in passage through the world, rather than stated explicitly. Lorde, "Age, Race, Class, and Sex: Women Redefining Difference," in *Sister Outsider: Essays and Speeches* ((Berkeley, CA: Feminist Press, 1984) 114–23., 1984).

90. Nicholas Robinson, *The Christian Philosopher; or, A Divine Essay on the Doctrines of Man's Universal Redemption* (London, 1741) 141.

91. Genesis 2:7; 2:21.

92. 1 Timothy 2:13–15.

93. "Extraction, n.," *Oxford English Dictionary* (Oxford: Oxford University Press, March 2024) https://doi.org/10.1093/OED/1008674222. Also see Samuel Johnson's *Dictionary*'s definition of "extraction, n.s.": "1. The act of drawing one part out of a compound; the act of drawing out the principal substance by chemical operation." *A Dictionary of the English Language*, by Samuel Johnson (1755) https://johnsonsdictionaryonline.com/views/search.php?term=extraction (accessed July 6, 2022).

94. James Grantham Turner, *One Flesh: Paradisal Marriage and Sexual Relations in the Age of Milton* (Oxford: Clarendon, 1987) 26.

95. Ibid. 100.

96. Ibid. 16.

97. Donna Landry, *The Muses of Resistance: Laboring-Class Women's Poetry in Britain, 1739–1796* (Cambridge: Cambridge University Press, 1990) 57.

98. Goodridge offers a precise account of the ways in which Collier adds back in women's work (*Rural Life*, see esp. 20, 33, 66–67).

99. Landry resolves the problem by pushing Collier's argument toward textuality: she "gives us a textualization of women's work as social and material, but not exclusively or primarily biological, reproduction" (*Muses of Resistance* 57). I am suggesting, by contrast, that Collier keeps the material body in play, even as she takes control of defining its material difference.

100. Iris Marion Young, "Throwing like a Girl: A Phenomenology of Feminine Body Comportment Motility and Spatiality," *Human Studies* 3.2 (1980): 141, 140.

101. Ibid. 147, 148.

102. Ibid.

103. Gayle Salamon, "The Phenomenology of Rheumatology: Disability, Merleau-Ponty, and the Fallacy of Maximal Grip," *Hypatia* 27.2 (2012): 244.

104. Ibid. 253.

105. Ibid. 245.

106. Jonathan Swift, "The Progress of Beauty," *Poems of Jonathan Swift*, vol. 1, ed. Harold Williams (Oxford: Oxford University Press, 1958) ll. 45, 75.

107. Tita Chico, "Privacy and Speculation in Early Eighteenth-Century Britain," *Cultural Critique* 52 (2002): 41.

108. Ibid. 45.

109. Jonathan Kramnick, *Paper Minds: Literature and the Ecology of Consciousness* (Chicago: University of Chicago Press, 2018) 86.

110. Swift, "Progress" ll. 58–59.

111. Kramnick, *Paper Minds* 87. Fairer's model of eco-georgic as a "functioning system" underlies Kramnick's conception of what it might mean for the woman's face to decay into the landscape. This model for the georgic stands to naturalize power relations that exist in the society from which the poem and its models come ("'Where Fuming Trees'" 205).

112. Chloe Wigston Smith, *Women, Work, and Clothes in the Eighteenth-Century Novel* (Cambridge: Cambridge University Press, 2013) 1.

113. Laura Brown, *Ends of Empire: Women and Ideology in Early Eighteenth-Century English Literature* (Ithaca, NY: Cornell University Press, 1993) 43. Brown describes the commodities in *Oroonoko* this way, but in this reading of Collier I have in mind her reading of Pope's Belinda, chap. 4.

114. Janet Lembke, introduction to *Virgil's Georgics: A New Verse Translation*, trans. Janet Lembke (New Haven, CT: Yale University Press, 2005) xiii.

115. Fairer, "Georgic" 463.

116. Fairer, "'Where Fuming Trees'" 205.

117. Carol Gilligan, *In a Different Voice: Psychological Theory and Women's Development* (Cambridge, MA: Harvard University Press, 1982). Gilligan describes the gap between "a morality of rights that dissolves 'natural bonds' in support of individual claims and a morality of responsibility that knits such claims into a fabric of relationship, blurring the distinction between self and other through the representation of their interdependence" (132). She champions an ethics centered on "responsibility" and "activities of care" (132).

118. Sennett, *Craftsman* 8.

119. Ibid. 289.

Chapter 3

I developed the ideas in this chapter in an earlier form in "Hogarth's Practical Aesthetics," in *Mind, Body, Motion, Matter*, ed. Mary Helen McMurran and Alison Conway (Toronto: University of Toronto Press, 2016) 21–46.

1. Ephraim Chambers, *Cyclopaedia; or, an Universal Dictionary of Arts and Sciences* (London, 1728) ii.

2. Larry Shiner, *The Invention of Art: A Cultural History* (Chicago: University of Chicago Press, 2001) 81. Also see Peter Greenhalgh, who explains that the "fine arts" were formulated in the sixteenth century but systematized in the eighteenth ("The History of Craft," *The Culture of Craft: Status and Future*, ed. Peter Dormer [Manchester: Manchester University Press, 1997] 20–52).

3. For the larger argument, see Shiner, *Invention of Art* chap. 5. For an expansion and critique of Shiner's argument, see Glenn Adamson, *The Invention of Craft* (London: Bloomsbury, 2013). Adamson argues that Shiner overstates the separation of craft from fine art, defining the latter as "a largely twentieth-century, even post-1945, tendency" (ibid. xiv). Adamson conceives of eighteenth-century craft positioned not against art but against industrialization" (ibid. xix).

4. William Hogarth, *The Analysis of Beauty* (ed. Ronald Paulson [New Haven, CT: Yale University Press, 1997]). Hereafter cited as *AB*.

5. See Celina Fox's account of Hogarth's disillusionment with the Society and his departure in 1757 (*Arts of Industry in the Age of Enlightenment* [New Haven, CT: Yale University Press, 2009] 207).

6. Celina Fox traces the society's Baconian conception to Philip Peck's letter to the Royal Society in 1738, though that Society would not take up the project. In Fox's words,

the Royal Society "had no wish to serve as a joint-stock company and exploit invention for profit" (ibid. 184).

7. Ibid. 186.

8. Ibid.

9. Henry Baker, quoted in ibid. 187.

10. Jenny Uglow, *Hogarth: A Life and a World* (New York: Farrar, Straus & Giroux, 1997) 31–32.

11. In one of the unpublished manuscripts of the *Analysis,* Hogarth explains his desire to be "an History Painter or engraver" after "having in the beginning of life lost a great part of my time, in engraving coats of armes on Silver Plate." He does not mention this history in the published text ("Manuscripts of *The Analysis of Beauty,*" in Paulson, ed., *Analysis of Beauty* 121). Moreover, Hogarth would disparage the drudgery of "mechanical copying" and in many cases separate out his identity as an "artist" from the early instruction of "a master who was fundamentally an artisan" (Ronald Paulson, *Hogarth,* 3 vols. [New Brunswick, NJ: Rutgers University Press, 1991–93] 1:50, 1:47).

12. Howard Risatti offers the establishment of separate artists and engravers as an early example of the splitting off of craft from art (*A Theory of Craft: Function and Aesthetic Expression* [Chapel Hill: University of North Carolina Press, 2007] 103–4).

13. Paulson, *Hogarth* 3:xv. Also see Paulson's account of Hogarth's fascination with "demotic art," like line drawings and stick figures portrayed in his engravings (3:123).

14. Uglow, *Hogarth* 518.

15. As Charles Saumarez Smith relates, Thornhill "became interested in the surrounding visual context in which his paintings would appear" (*Eighteenth-Century Decoration: Design and the Domestic Interior in England* [New York: Harry N. Abrams, 1993] 41). Hogarth was very much in touch with such a vision of the decorated interior, with his own prints used as "furniture" or decoration. See Diana Donald, "'The Truly Natural and Faithful Painter': Hogarth's Depiction of Modern Life," in *Hogarth: Representing Nature's Machines,* ed. David Bindman, Frédéric Ogée, and Peter Wagner (Manchester: Manchester University Press, 2001) 166.

16. Paulson, *Hogarth* 3:11.

17. Sir Joshua Reynolds, *Discourses on Art* (New Haven, CT: Yale University Press, 1997) 19, 18. See John Barrell, *Political Theory of Painting from Reynolds to Hazlitt: "The Body of the Public"* (New Haven, CT: Yale University Press, 1986) chap. 1. This is Reynold's Royal Academy rhetoric. For an account of his practice as deeply engaged with natural philosophy, see Matthew C. Hunter, *Painting with Fire: Sir Joshua Reynolds, Photography, and the Temporally Evolving Chemical Object* (Chicago: University of Chicago Press, 2019).

18. Barrell, *Political Theory* 79.

19. Paulson, *Hogarth* 3:11; internal quote is from Nikolaus Pevsner, *Academies of Art Past and Present* (Cambridge: Cambridge University Press, 1940).

20. The problem of attaching the *Analysis* to Hogarth's visual art is an old one. See Michael Baridon, "Hogarth's 'Living Machines of Nature' and the Theorisation of Aesthetics," in Bindman, Ogée, and Wagner, eds., *Hogarth* 81. Abigail Zitin makes the claim for necessarily separating the two in her *Practical Form: Abstraction, Technique, and Beauty*

in *Eighteenth-Century Aesthetics* (New Haven, CT: Yale University Press, 2020) 106: "I maintain that Hogarth's theory resists easy integration with, or application to, his graphic works." This drastic separation is result of Zitin's dedication to pulling Hogarth into a process-only aesthetics in line with the nonteleological character of Kant's "pure beauty." The result is strange: Zitin argues that Hogarth embraces an "artisanal" "phenomenology" but she cannot discuss any of the artifacts (material ends) created by the artisan, including the two explanatory plates that were included with his written treatise.

21. Paulson, *Hogarth* 3:63. Uglow connects Hogarth to the world of public scientists and experimenters (*Hogarth* 520).

22. For connections between Hogarth's art and experimental philosophy, see Frédéric Ogée's "Je-Sais-Quoi: William Hogarth and the Representation of the Forms of Life," where he reports on Hogarth's engravings evidencing "new ways of observing nature" and analyzing the "experiments" they represent "in a very scientific manner" (in Bindman, Ogée, and Wagner, eds., *Hogarth* 71–72).

23. Francis Bacon, *The New Organon* (ed. Lisa Jardine and Michael Silverthorne [Cambridge: Cambridge University Press, 2000]) 18, 21.

24. In an earlier draft of the introduction, Hogarth uses the phrase "Power of habit and custom," which is closer to Locke's account in the *Essay Concerning Human Understanding* ("Manuscripts" 115).

25. Donald, "Truly Natural" 87.

26. See Ronald Paulson, introduction to Hogarth, *Analysis* xxxx.

27. Critics have never had a doubt that Hogarth was writing for both audiences. On Hogarth's relation to Shaftesbury and Hutcheson see Ronald Paulson, *The Beautiful, Novel, and Strange: Aesthetics and Heterodoxy* (Baltimore: Johns Hopkins University Press, 1996) esp. ch. 3.

28. George Dickie, *The Century of Taste: The Philosophical Odyssey of Taste in the Eighteenth Century* (New York: Oxford University Press, 1996) 11.

29. Ibid.

30. Joseph Addison, "The Pleasures of the Imagination," *The Spectator*, ed. Donald F. Bond (Oxford: Clarendon, 1965) 3:537. Bond, for his part, attaches this claim to an "old tradition" of writing on imagination (Addison, "Pleasures" 3:537, n3).

31. Both Paul Guyer and Robert R. Clewis end their discussions in Kant's "adherent beauty," a concept-driven beauty, which they argue should be taken more seriously than it has been. Guyer, "Beauty and Utility in Eighteenth-Century Aesthetics," *Eighteenth-Century Studies* 35.3 (2002): 439-53; Clewis, "Beauty and Utility in Kant's Aesthetics: The Origins of Adherent Beauty," *Journal of the History of Philosophy* 56.2 (2018): 305–35.

32. Francis Hutcheson, *An Inquiry into the Original of Our Ideas of Beauty and Virtue*, 4th ed. (London, 1738) 11; qtd. in Paul Guyer, "Beauty and Utility" 442.

33. George Berkeley, *Alciphron, or the Minute Philosopher* in *The Works of George Berkeley, Bishop of Cloyne*, ed. A. A. Luce and T. E. Jessop (London: Nelson, 1950) 3:124; qtd. in Guyer, "Beauty and Utility" 441. My outline of this discussion is indebted to Guyer (who does not include Hogarth among his eighteenth-century aesthetic philosophers).

34. Hutcheson, *Inquiry* 304; qtd. in Guyer, "Beauty and Utility" 442.

35. See J. T. A. Burke, "A Classical Aspect of Hogarth's Theory of Art," *Journal of the Warburg and Courtauld Institutes* 6.1 (1943): 151–53.

36. Guyer, *"Beauty and Utility"* 443.

37. Although this chapter receives frequent brief reference, it is hard to come by a full reading of it. Even Jules Lubbock in his book on taste and design characterizes it as posing "considerable difficulty" for the reader in "working out exactly what Hogarth is trying to say." He continues, "My belief, arrived at after discussing the passage with many puzzled students over many years is that Hogarth wrote it deliberately as a mind-bending riddle in order to convey his notion that the relationship between fitness and beauty was an extremely intricate one, a prime example, therefore of intricacy which he considered the crucial component of beauty" (*The Tyranny of Taste: The Politics of Architecture and Design in Britain 1550–1960* [New Haven, CT: Yale University Press, 1995] 195).

38. Xenophon, *Memorabilia* (*Memorabilia. Oeconomicus. Symposium. Apology*, trans. E. C. Marchant and O. J. Todd [Cambridge, MA: Harvard University Press, 2014] 233).

39. Ibid.

40. Ibid. 251.

41. Ibid.

42. Ibid.

43. Ibid. 253.

44. Risatti describes the craft object as having a "conceptual approach to the body" that may be understood in terms of "containing, covering, and supporting" (*Theory of Craft* 18). For him, then, the craft object is always in some way for the body, and its "intended function" is built into it. "This is why an empty chair or empty pot still exhibit their intended function" (27, 23). An empty chair leaves "a negative imprint of the human body" (109).

45. Ibid. 111–12.

46. Ibid. 19.

47. Ibid. 17.

48. Ibid. 27.

49. Ibid. chap. 8.

50. "Material, *N.*, Sense 2.a.i," *Oxford English Dictionary* (Oxford: Oxford University Press, March 2024) https://doi.org/10.1093/OED/4237015700.

51. "Throw, *V.* (1), Sense II.9," *Oxford English Dictionary* (Oxford: Oxford University Press, June 2024) https://doi.org/10.1093/OED/1075609129.

52. "Throw, *V.* (1), Sense II.6.b," *Oxford English Dictionary* https://doi.org/10.1093/OED/2018313069.

53. Hogarth, "Manuscripts" 119.

54. Ibid.

55. Tim Ingold describes engraving as having "broke[n] the link between the gesture and its trace" (*Lines: A Brief History* [London: Routledge, 2016] 142).

56. Paulson, *Hogarth* 3:101.

57. Jean Francois Billeter, *The Chinese Art of Writing* (New York: Rizzoli, 1990) 47; qtd. in Ingold, *Lines* 50.

58. Reviel Netz, *The Shaping of Deduction in Greek Mathematics: A Study in Cognitive History* (Cambridge: Cambridge University Press, 1999) 12.

59. Ibid. 33.
60. Ibid. 35.
61. Ibid. 54, 56.
62. Ibid. 56.
63. Ibid. 57.
64. Ibid. 60. In the final, historical narrative of the book, Netz remarks that Greek mathematicians may have "felt uneasily close to the banausic" (ibid. 61), and he describes the "mechanical" nature of drawn diagram, its place in the "material world" revealing an "estrangement between the theoretical and the practical" (ibid. 303).

65. I consider Hogarth in terms of the history of design in "Hogarth's Practical Aesthetics," in *Mind, Body, Motion, Matter*, ed. Mary Helen McMurran and Alison Conway (Toronto: University of Toronto Press, 2016) 21–46.

66. Uglow, *Hogarth* 525.

67. Paulson, introduction xlvi. This repeats in stronger language Paulson's claim in *Hogarth* 3:125.

68. Uglow, *Hogarth* 540.

69. See Paulson, *Hogarth* 3:58.

70. The eighteenth century saw a new scientific interest in lifelike presentation of specimens. See Valérie Kobi, "Staging Life: Natural History Tableaux in Eighteenth-Century Europe," *Journal18* 3 (2017): www.journal18.org/1306.

71. Ingold, *Lines* xv.

72. Ibid. 1.

73. E. H. Gombrich, *Art and Illusion: A Study in the Psychology of Pictorial Representation* (Princeton, NJ: Princeton University Press, 1960) 7.

74. See Paulson's example of Hogarth's line art in *Hogarth* 3:123.

75. Gombrich, *Art and Illusion* 306.

76. Ibid.

77. Ibid.

78. Jonathan Kramnick, *Paper Minds: Literature and the Ecology of Consciousness* (Chicago: University of Chicago Press, 2018) 91.

79. Ibid.

80. John Bender's analysis of Hogarth's graphic works in terms of a self-reflexive "facticity" draws out how the prints "call attention to the means of illusion (perspective, for instance) and tend to fly to the pictorial surface in collage-like ways" (*Ends of Enlightenment* [Stanford, CA: Stanford University Press, 2012] 59.

81. Uglow, *Hogarth* 532.

82. Leon Battista Alberti, *Of Painting* (trans. James Leoni, [London, 1755]) 241.

83. Ingold, *Lines* 42. Ingold quotes Alberti at the beginning of his chapter on "Traces, Threads, and Surfaces," though he does not analyze the passage in any detail. Like

Hogarth, however, he is interested in disentangling what Alberti seems to force together: the difference between surface, line, and thread.

84. Robert Briscoe, "Gombrich and the Duck-Rabbit," *Aspect Perception after Wittgenstein: Seeing-As and Novelty*, ed. Michael Beaney, Brendan Harrington, and Dominic Shaw (New York: Routledge, 2018) 50.

85. J. J. Gibson, *The Ecological Approach to Visual Perception* (Boston: Houghton Mifflin, 1979) 281; qtd. in Briscoe, "Gombrich" 51.

86. Briscoe, "Gombrich" 53–54.

87. Gombrich, *Art and Illusion* 6.

88. Briscoe, "Gombrich" 49.

89. David Bindman relates that Reynolds called out Hogarth in "The Idler" papers for a "'servile attention to minute exactness' [which was] wholly inimical to true grandeur" (*Hogarth and His Times: Serious Comedy* [Berkeley: University of California Press, 1998] 15). Reynolds would have known that Hogarth had already responded to other such allegations by attempting to distance himself with a "hilariously coarse burlesque etching in 'the true Dutch taste' . . . of his own . . . painting" (Bindman, *Hogarth and His Times* 17). The connection was not only in the detail. Like the Dutch artists Alpers considers, Hogarth works against an Italian model of linear perspective; Paulson describes his "modern moral progresses" as including such a varied set of perspectives as to seem to the viewer "shaky" (*Hogarth* 3:61). And like those artists, Hogarth, as I am suggesting, is quite concerned with the "surface" of the painting and its relation to the craftsman. See Svetlana Alpers, *The Art of Describing: Dutch Art in the Seventeenth Century* (Chicago: University of Chicago Press, 1983) xxiv.

90. We might understand the relation between the decorative frame and the how-to craftwork by way of Adamson's account, which turns back to Derrida on painting to argue that decoration and craft share a similar kind of supplemental relation to the work of art (*Invention of Art* 12).

91. Joshua Kirby, *Dr. Brook Taylor's Method of Perspective Made Easy, Both in Theory and Practice* (London, 1754). The title page offers a claim to expertise similar to Hogarth's: "by Joshua Kirby, Painter." Craig Ashley Hanson's reading of Hogarth may be the most sympathetic to this view of the *Analysis*. Contextualizing Hogarth in terms of the "English virtuoso," Hanson explains that the plate "calls to mind the Royal Society History of Trades program," and he also suggests a link between Royal Society collecting and the plate, noting Hogarth's "polymathic pursuits" and "interest in shells and flowers" (*The English Virtuoso: Art, Medicine, and Antiquarianism in the Age of Empiricism* [Chicago: University of Chicago Press, 2009] 155).

92. Paulson, *Hogarth* 3:107.

93. James Noggle, *The Temporality of Taste in Eighteenth-Century British Writing* (Oxford: Oxford University Press, 2012) 1.

94. Ibid. 3.

95. Michel de Certeau, *The Practice of Everyday Life*, trans. Steven Rendall (Berkeley: University of California Press, 1984) xviii.

96. Ibid. xviii. Deligny began his project of exploring autistic life beyond the institution in 1967 in the Cevennes at Monoblet. The children were between three and ten and

were mute, thus offering the possibility to see experience organized by those "who cannot speak the dominant language" (Erin Manning, *Always More than One: Individuation's Dance* [Durham, NC: Duke University Press, 2013] 190).

97. De Certeau, *Practice* xix.

98. Pierre Bourdieu, *Outline of a Theory of Practice*, trans. Richard Nice (Cambridge: Cambridge University Press, 1977) 2.

99. Ibid. 2.

100. Ibid. 38. Tim Ingold distinguishes between two kinds of trace lines: "*trace of a gesture*" and "an assembly of *point-to-point connectors*" (77). Ingold reaches back to the British eighteenth century for a clear example of his own culture's investment in the "trace" of a "gesture," taking as his neat visual example Sterne's depiction of Corporal Trim's walking stick in *Tristram Shandy* (74–75). He associates this figure with the "walk" (as in Hogarth) and with the "wayfar[er]" (88).

101. See Michael Snodin, "Style," in *Design and the Decorative Arts: Britain, 1500–1900*, ed. Michael Snodin and John Styles (London: V & A, 2001) 44–46; Paulson, *Hogarth* 3: 123. While Paulson claims that the serpentine line was meant to serve a broader "normative function," he acknowledges that "contemporaries did, associate the serpentine line with rococo forms" (*Hogarth* 3:122).

102. Snodin, "Style" 44, 46. Snodin describes rococo as a "style without rules" and the *Analysis* as "the nearest rococo ever came to a theoretical justification" (*Design* 44).

Chapter 4

1. As "ethnographic" as Banks may seem to be, there of course was no scientific discipline by this name in the 1760s. Not only that, but, as Adrienne Kaeppler explains, "it was not until decades later that the social and cultural history of man was considered, except by philosophers, worthy of serious study" (*"Artificial Curiosities": An Exposition of Native Manufactures Collected on the Three Pacific Voyages of Captain James Cook, R.N.* [Honolulu: Bernice Pauahi Museum, 1978] 37).

2. J. C. Beaglehole, introduction to *The Endeavour Journal of Joseph Banks*, 2 vols., ed. J. C. Beaglehole (Sydney: Public Library of New South Wales, 1963) 1:40. Hereafter cited in the text as B.

3. Margaret Cohen, *The Novel and the Sea* (Princeton, NJ: Princeton University Press, 2010) 42–44.

4. The copy belonged to Sydney Parkinson, Banks's draftsman. Parkinson lists Hogarth's title in his sketchbook, so he likely had his own copy on board. See Bernard Smith, *Imagining the Pacific in the Wake of the Cook Voyages* (New Haven, CT: Yale University Press, 1992) 87. Smith discusses Hogarth's influence on Parkinson's work on the voyage (87–89).

5. At least within writing on Oceania, this qualification is usually attributed to Nicholas Thomas. See, for example, his claim that "colonial ideologies may have been more variable, complex, and ambivalent than has been generally acknowledged" (*Colonialism's Culture: Anthropology, Travel, and Government* [Princeton, NJ: Princeton University Press, 1994] 17). For the use of this concept in analysis of travel writing, see Jonathan Lamb's description of the "messy" imperial eye (*Preserving the Self in the South Seas, 1680–1840*

[Chicago: University of Chicago Press, 2001] 7); Nigel Leask's account of the "*vulnerability . . .* of European travelers" (*Curiosity and the Aesthetics of Travel Writing, 1770–1840* [New York: Oxford University Press, 2002] 16); and Paul Smethurst's account of "vulnerability and confusion" in the accounts of the period (*Travel Writing and the Natural World, 1768–1840* [New York: Palgrave, 2012] 64).

6. In the journal, Banks, following Cook, calls these "Bow Island," "the groups," and "Bird Island" (*B* 1:245–48).

7. *The Journals of Captain James Cook*, vol. 1, *The Voyage of the* Endeavour, *1768–1771* (ed. J. C. Beaglehole [Cambridge: Hakluyt Society and Cambridge University Press, 1955] 71. Hereafter cited in the text as C.

8. William Dampier, *Memoirs of a Buccaneer: Dampier's New Voyage Round the World, 1697* (Mineola, NY: Dover, 1968) 4.

9. Philip Edwards, *The Story of the Voyage: Sea-Narratives in Eighteenth-Century England*, (Cambridge: Cambridge University Press, 1994) 27. Edwards calls Dampier a "natural Baconian scientist" (20). For Edwards's full account of Dampier and the Royal Society, see 27–40.

10. Smethurst, *Travel Writing* 22.

11. See Daniel Carey, "Inquiries, Heads, and Directions: Orienting Early Modern Travel," *Travel Narratives, the New Science, and Literary Discourse, 1569–1750*, ed. Judy A. Hayden, (Burlington, VT: Ashgate, 2012) 26.

12. Jason H. Pearl, "Geography and Authority in the Royal Society's Instructions for Travelers," *Travel Narratives, the New Science, and Literary Discourse, 1569–1750*, ed. Judy A. Hayden, (Burlington, VT: Ashgate, 2012) 74. In terms that resonate with some of my conclusions later in the chapter, Pearl sees these conventions as "stripping away human connection" from the reports. For a history of objectivity that highlights the way that scientific instruments remove the body from experience, see Joanna Picciotto, *Labors of Innocence in Early Modern England* (Cambridge, MA: Harvard University Press, 2010).

13. On the astronomical results from the voyage, see Wayne Orchiston, who claims (contra Beaglehole) that the observations of the Transit made an important contribution to science ("From the South Seas to the Sun: The Astronomy of Cook's Voyages," *Science and Exploration in the Pacific: European Voyages to the Southern Oceans in the Eighteenth Century*, ed. Margarette Lincoln [Rochester, NY: Boydell Press and National Maritime Museum, 1998] 55–72).

14. R. A. Skelton, "The Graphic Records," *C* 1:cclxv.

15. For a record of the many surviving charts see Skelton, Ibid. cclxv-cclxxi.

16. Bruno Latour, "Visualization and Cognition: Thinking with Eyes and Hands," *Knowledge and Society Studies in the Sociology of Culture Past and Present: A Research Annual* 6 (1986) 7, 3, 3, 3.

17. Ibid. 15.

18. Although these drawings were not as immediately engraved and circulated as we would expect, they laid the foundation for Banks's botanical imperialism. See John Gascoigne, "Joseph Banks and the Expansion of Empire," in Lincoln, ed., *Science and Exploration in the Pacific* 39–54.

19. Cohen, *Novel and the Sea* 44.

20. William Eamon has charted this history in *Science and the Secrets of Nature: Books of Secrets in Medieval Culture* (Princeton, NJ: Princeton University Press, 1996). See particularly his account of how Bacon's "experimental histories of the trades" take the form of the recipe (7).

21. Ibid. 7.

22. Gilbert Ryle, "Knowing How and Knowing That: The Presidential Address," *Proceedings of the Aristotelian Society* n.s. 46 (1945–46): 1–16.

23. Beaglehole, introduction to *C* 1: cix.

24. Without exactly putting it this way, biographers have long concentrated on Banks's body as a source of his distinctive reporting. See Patrick O'Brian's account of Banks's "superabundant energy" (*Joseph Banks: A Life* [Boston: David R. Godine, 1993] 90); and Neil Rennie's account of his "temperament" (*Far-Fetched Facts: The Literature of Travel and the Idea of the South Seas* [Oxford: Clarendon, 1995] 91).

25. Cohen, *Novel and the Sea* 57.

26. Ibid. 43.

27. Beaglehole calls Cook "the genius of the matter-of-fact" (*C* 1:cxcii) and describes his "workmanlike prose" this way: "He regards words unromantically, as concrete things with a precise use, much in the way he regards a block and tackle or a tiller-brace" (*C* 1:cxciii). This conception of Cook's language is intriguingly Baconian.

28. Qtd. in John Gascoigne, *Joseph Banks and the English Enlightenment: Useful Knowledge and Polite Culture* (Cambridge: Cambridge University Press, 1994) 138.

29. Michel Foucault, *The Order of Things: An Archaeology of the Human Sciences* (New York: Routledge, 2002) 82.

30. Ibid. 74.

31. Mary Louise Pratt, *Imperial Eyes: Travel Writing and Transculturation* (New York: Routledge, 1992) 31.

32. Ibid. 35.

33. Ibid. 31, 37.

34. Foucault, *Order of Things* 74.

35. Ibid. 131.

36. Ibid. 130.

37. Lorraine Daston and Peter Galison, *Objectivity* (New York: Zone, 2007) 81.

38. Foucault, *Order of Things* 132.

39. Ibid. 133.

40. Ibid.

41. Ibid. 132.

42. Daston and Galison, *Objectivity* 82.

43. Both Rennie and Vanessa Smith use the term to describe Banks's participation in the mourning ceremony (Rennie, *Far-Fetched Facts* 91; Smith, "Performance Anxieties: Grief and Theatre in European Writings on Tahiti," *Eighteenth-Century Studies* 41.2 [2008]: 155).

44. I follow the anthropologist Barbara Tedlock who calls the term an "oxymoron," even as she acknowledges that it is the "principal mode of production for anthropological knowledge" ("From Participant Observation to the Observation of

Participation: The Emergence of Narrative Ethnography," *Journal of Anthropological Research* 47.1 [1991]: 69).

45. Douglas L. Oliver, *Ancient Tahitian Society* (Honolulu: University of Hawaii Press, 1974) 502, 526.

46. Ibid. 503.

47. Ibid. 505.

48. See *B* 288, 4n.

49. Smith, "Performance" 156.

50. Ibid.

51. *Character* is of course a complex term in this period. In beginning with a separation between actor/role or body/clothes, I draw on Vanessa Smith's assumption that Banks plays into a surface/depth model she associates with the theatricality of sympathy. Banks's materialist, surface conception of character, however, may be closer to what Lisa Freeman defines for "theater's character": "a tenacious emphasis on readable surfaces and .. resistance to notions of "Sincere" depths (*Character's Theater: Genre and Identity on the Eighteenth-Century English Stage* [Philadelphia: University of Pennsylvania Press, 2002] 39).

52. "To mark with some black or dirty substance; to blacken, smudge" ("Smut, *V.*, Sense 1.a," *Oxford English Dictionary* [Oxford: Oxford University Press, December 2023] https://doi.org/10.1093/OED/4289456140).

53. Lamb describes a voyaging early modern self "in isolation from social structure" (*Preserving the Self* 19). We might understand Banks to comprehend that there is no grasping "social structure"—his own or another's—without relinquishing what Lamb sees as a requirement for the vulnerable European traveler: "self-preservation."

54. "To lay or embed (a thing) in the substance of something else so that its surface becomes even or continuous with that of the matrix" ("Inlay, *V.*, Sense 2.a," *Oxford English Dictionary* (Oxford: Oxford University Press [June 2024] https://doi.org/10.1093/OED/7427370384).

55. Alfred Gell, *Wrapping in Images: Tattooing in Polynesia* (Oxford: Clarendon, 1996) 39, 38.

56. Ibid. 39.

57. Nicholas Thomas, *In Oceania: Visions, Artifacts, Histories* (Durham, NC: Duke University Press, 1997) 4.

58. David Turnbull, "Reframing Science and Other Local Knowledge Traditions," *Futures* 29.6 [1997]: 556. Turnbull describes the map as a "knowledge assemblage" ("Reframing Science," 553). Elsewhere, Turnbull lays out the difference between knowledge systems as one between "representationalism" and "performativity" ("Cook and Tupaia, a Tale of Cartographic *Méconnaissance*?" in Lincoln, ed., *Science and Exploration in the Pacific* 131).

59. Turnbull, "Reframing Science" 556.

60. Ibid.

61. Lars Eckstein and Anja Schwarz, "The Making of Tupaia's Map: A Story of the Extent and Mastery of Polynesian Navigation, Competing Systems of Wayfinding on James Cook's *Endeavour*, and the Invention of an Ingenious Cartographic System," *Journal of Pacific History* 54.1 (2019): 1–95. The authors summarize the "draft stages" on 16–18.

62. Ibid. 32.
63. Ibid.
64. Ibid.
65. For a detailed account of Banks's involvement in Tupaia's map, also see Harriet Parsons, "British-Tahitian Collaborative Drawing Strategies on Cook's Endeavour Voyage," in *Indigenous Intermediaries: New Perspectives on Exploration Archives*, ed. Shino Konishi, Maria Nugent, and Tiffany Shellam (Canberra: ANU Press, 2015) 155–56.
66. For an example of a marginal sketch that seems in line with what Banks offers here, see Daston and Galison's reproduction of Rene-Antoine Ferchault de Reaumur's drawing of insect antennae (*Objectivity* 87). Linnaeus was a notoriously bad draftsman; see Isabelle Charmantier, "Carl Linnaeus and the Visual Representation of Nature," *Historical Studies in the Natural Sciences* 41.4(2011): 365–404.
67. Barbara Maria Stafford describes the contribution of the illustrations as the achievement of "semiotic wholeness" (*Voyage into Substance: Art, Science, Nature, and the Illustrated Travel Account, 1760–1840* [Cambridge, MA: MIT Press, 1984] 51). More than those lavish illustrations, Banks's sketches may recall the small engravings from Dampier's famous text of the 1660s, which have been associated with Dampier's "sober, circumstantial style" and with the "fact" that style aimed to project (Lamb, *Preserving the Self* 59), though they take on a different role here.
68. Jasamin Kashanipour, "The Gradual Gaze: Drawing as a Practice of Ethnographic Description," *Anthropology and Humanism* 46.1 (2021) 81. See also Michael Taussig's account of his own ethnographic drawings in *I Swear I Saw This: Drawings in Fieldwork Notebooks, Namely My Own* (Chicago: University of Chicago Press, 2011). Notice how Taussig's description of his project draws out a familiar metaphor from Banks's journal: "To draw is to apply pen to paper. But to draw is also to pull on some thread, pulling it out of its knotted tangle or skein, and we also speak of drawing water from a well. There is another meaning too, as when we say 'I was drawn to him' . . . Drawing is thus a depicting, a hauling, an unraveling, and being impelled toward something or somebody" (*I Swear* xii).
69. *Objectivity* 86.
70. Kashanipour, "Gradual Gaze" 81.
71. Despite the presence of the *Analysis of Beauty* onboard the *Endeavour* (brought by Parkinson), readers of Banks have perpetually turned to Reynolds as a way to explain the journal's treatment of aesthetics. See Harriet Guest, "Curiously Marked: Tattooing and Gender Difference in Eighteenth-Century British Perceptions of the South Pacific," *Written on the Body: The Tattoo in European and American History*, ed. Jane Caplan (Princeton, NJ: Princeton University Press, 2000) 85; and Smethurst, *Travel Writing* 9.
72. Michael Taussig, *Mimesis and Alterity: A Particular History of the Senses* (London: Routledge, 1993) 20–21.
73. Thomas, *In Oceania* 10, 106.
74. Guest, "Curiously Marked" 85.
75. Timothy Ingold, *Lines: A Brief History* (New York: Routledge, 2016) 1–5.
76. Paul Carter, *The Road to Botany Bay: An Exploration of Landscape and History* (New York: Alfred A. Knopf, 1988) xv.
77. Ibid. xxii.

78. Ibid.

79. Horkheimer and Adorno's account of violent Reason still remains influential, if only in the kinds of knowledge that are privileged in order to combat it.

80. Richard Sennett, *The Craftsman* (New Haven, CT: Yale University Press, 2008) 11.

81. On the violence of Bacon's metaphors, see Carolyn Merchant, "'The Violence of Impediments': Francis Bacon and the Origins of Experimentation," *Isis* 99.4 (2008): 731–60.

Chapter 5

1. James Boswell, *The Journal of a Tour to the Hebrides* (*Boswell's Life of Johnson*, vol. 5, ed. George Birkbeck Hill [Oxford: Clarendon, 1964] 77). Hereafter cited in the text as JB.

2. Samuel Johnson does not mention this exchange at all and barely mentions seeing Monboddo (*A Journey to the Western Islands of Scotland* (ed. Mary Lascelles [New Haven, CT: Yale University Press, 1971] 12). Hereafter cited in the text as SJ.

3. "Sage" (JB 55, 358, 370); "philosopher" (JB 246, 261, 344). For instance, Peter Martin (one of Boswell's biographers) in just this vein calls Johnson "the high priest of late-century neoclassicism and rationality" (*A Life of James Boswell* [New Haven, CT: Yale University Press, 2000] 300).

4. Pat Rogers, "Introduction: The Grand Detour of Johnson and Boswell," *Johnson and Boswell in Scotland: A Journey to the Hebrides*, ed. Pat Rogers (New Haven, CT: Yale University Press, 1993) x. Rogers offers this as a paraphrase of the poet Anna Seward's comment on the two books in a letter to Boswell on the publication of the Tour: "In *one* we perceive, through a medium of solemn and sublime eloquence, in what light Scotland, her nobles, her professors, and her chieftains appeared to the august wanderer; in the *other* how the growling philosopher appeared to them" (qtd. Rogers, "Introduction" x). Prominent reiterations of the claim may be found in Peter Martin (*Life* 303) and in Paul Korshin, "'Extensive View': Johnson and Boswell as Travelers and Observers," *All Before Them: 1660–1780*, ed. John McVeagh (London: Ashfield, 1990) 233.

5. Gordon Turnbull, "'Generous Attachment': The Politics of Biography in the *Tour to the Hebrides*," *Dr. Samuel Johnson and James Boswell*, ed. Harold Bloom (New York: Chelsea House, 1986) 227.

6. John B. Radner, *Johnson and Boswell: A Biography of Friendship* (New Haven, CT: Yale University Press, 2012) 116.

7. As Radner explains it, Johnson eventually suggested that Boswell might publish if Johnson could review the text beforehand. For this particular episode of "competition," see ibid. esp. 161–64.

8. There is a copious amount of criticism on Johnson's *Journey* and very little on Boswell's *Tour*. Even at this late date, Boswell continues to occupy the place of the biographer, rather than an independent voice in this exchange.

9. There is a long history of treating Johnson's *Journey* as "philosophical" and a strong sense that it should not be treated as ordinary travel writing but instead, as Mary Lascelles noted in 1965, it "must be taken on its own terms" ("Notions and Facts: Johnson and Boswell on Their Travels," *Johnson, Boswell, and Their Circle: Essays Presented to Lawrence Fitzroy Powell, in Honour of His Eighty-Fourth Birthday* [Oxford: Clarendon, 1965] 229).

In the case of Francis R. Hart, this means classing Johnson's with accounts of Montesquieu's and Goldsmith's fictional travelers ("Johnson as Philosophic Traveler: The Perfecting of an Idea," *ELH* 36.4 [1969]: 679–95). In more recent years, this has meant classing Johnson's account with the writings of Scottish Enlightenment philosophers: the most thoroughgoing accounts are in Mary Poovey, *A History of the Modern Fact: Problems of Knowledge in the Sciences of Wealth and Society* (Chicago: University of Chicago Press (1998) 249–64; and in Ian Duncan, "The Pathos of Abstraction: Adam Smith, Ossian, and Samuel Johnson," *Scotland and the Borders of Romanticism*, ed. Leith Davis, Ian Duncan, and Janet Sorensen (Cambridge: Cambridge University Press, 2004) 38–56.

10. Pat Rogers, *Johnson and Boswell: The Transit of Caledonia* (Oxford: Clarendon, 1995) 70. Rogers notes that Boswell had met Banks and Solander in 1771 and would search out Cook in 1776 (ibid. 70–71).

11. Richard Schwartz, *Samuel Johnson and the New Science* (Madison: University of Wisconsin Press, 1971) 43. Schwartz firmly attaches Johnson to the scientific world in part by stressing Johnson's scientific publications: from his early publication of the biography of Boerhaave, to his collaboration with Dr. Robert James on his *Medicinal Dictionary*, to his cataloging of the Harleian library. See esp. chap. 2.

12. Martin Martin, preface, to *A Description of the Western Islands of Scotland* (London, 1703) n.p.

13. Thomas Pennant, "An Account of Several Earthquakes Felt in Wales," *Philosophical Transactions* 71 (1781): 67–81.

14. Boswell had written to Robertson on April 15 of that year, soliciting from him a letter to Johnson to encourage the trip (Radner, *Johnson and Boswell* 111).

15. Qtd. in Ronald L. Meek, *Social Science and the Ignoble Savage* (New York: Cambridge University Press, 1976) 137.

16. For Smith's use of Charlevoix and Lafitau, see Meek, *Social Science* 121-25.

17. Ibid. 116ff. Johnson makes such attribution difficult, since, as Katie Trumpener remarks, he suppresses any mention of this intellectual context (*Bardic Nationalism: The Romantic Novel and the British Empire* [Princeton, NJ: Princeton University Press, 1997] 68). Claire Lamont points out that Boswell was part of the "generation of students" exposed to Smith's theory and that we know Johnson read Kames's *Historical Law-Tracts* (1758), which offers a similar account to Smith's ("Dr Johnson, the Scottish Highlander, and the Scottish Enlightenment," *Journal for Eighteenth-Century Studies* 12.1 [1989]: 52). Since Pat Rogers's claim that Johnson and Boswell's "concerns . . . lay [] with the science of man" and that "Johnson travelled to Scotland with a full sense of the renown which the Edinburgh school of thinkers and social critics had acquired," scholars have begun to consider Johnson's philosophical contribution in terms of this context (Rogers, *Johnson and Boswell* 2). See 249–64 and Duncan, "Pathos of Abstraction" 38.

18. Hugh Blair, *A Critical Dissertation on the Poems of Ossian, the Son of Fingal* (London, 1763) 1.

19. Martin, preface 207.

20. Ibid. 207, 207–8.

21. Ibid. 208.

22. Ibid. 207.

23. Sara Ahmed, *What's the Use? On the Uses of Use* (Durham, NC: Duke University Press, 2019) 25.

24. On the window as a technology for making such a claim (within and beyond Johnson's example), see Rachel Ramsey, "The Literary History of the Sash Window," *Eighteenth-Century Fiction* 22.2 (2009): 171–94.

25. Ralph E. Jenkins, "'And I travelled after him': Johnson and Pennant in Scotland," *Texas Studies in Literature and Language* 14.3 (1972): 445–62; Thomas Jemielity, "Thomas Pennant's Scottish *Tours* and *A Journey to the Western Islands of Scotland*," in *Fresh Reflections on Samuel Johnson: Essays in Criticism*, ed. Prem Nath (Troy, NY: Whitston, 1987) 312–27.

26. Gilbert Ryle, "Knowing How and Knowing That: The Presidential Address," *Proceedings of the Aristotelian Society* n.s. 46 (1945–46): 2.

27. Ibid. 8.

28. For more on the relation between Bacon and Johnson, see Schwartz, *Samuel Johnson* chap. 3.

29. Joseph Addison, *The Spectator*, ed. Donald F. Bond (Oxford: Oxford University Press, 1965) 3:540; Edmund Burke discusses "vastness of extent" in *A Philosophical Enquiry into the Origin of Our Ideas of the Sublime and Beautiful*, ed. James T. Boulton (Notre Dame, IN: University of Notre Dame Press, 1968) 72.

30. Paul Guyer, *A History of Modern Aesthetics*, vol. 1, *The Eighteenth Century* (Cambridge: Cambridge University Press, 2014) 64–65.

31. Abraham Cowley, "To the Royal Society," *The History of the Royal-Society of London, for the Improving of Natural Knowledge*, by Thomas Sprat (London, 1667) n.p.

32. Bradford Q. Boyd, "The Highland Tour Through the Spectacles of Books: Johnson, Pastoral, and Improvement in Late-Georgian Scotland," *Philological Quarterly* 100.3–4 (2021): 465, 469.

33. Ibid. 473.

34. Paul J. Alpers, *What Is Pastoral?* (Chicago: University of Chicago Press, 1996) 169.

35. For a different reading of this account of "romance," see Alison Hickey's account of this scene in terms of Johnson's "commanding vision" and the broader use of romance to appropriate Scotland ("'Extensive Views' in Johnson's Journey to the Western Islands of Scotland," *Studies in Eighteenth-Century Culture* 32.3 [1991]): 549, 551).

36. Elizabeth Fowler, "Art and Orientation," *New Literary History* 44.4 (2013): 598.

37. Ibid. 597.

38. Burke explains the sublime in terms of "astonishment" and "that state of the soul, in which all its motions are suspended." This prevents any attempt to "reason on that object which employs it" (*Enquiry* 57).

39. Joanna Picciotto, *Labors of Innocence in Early Modern England* (Cambridge, MA: Harvard University Press, 2010).

40. Johnson refers here to Hector Boece or Boethius (1465?–1536), the first principal of Aberdeen University and the author of *Scotorum Historiae* (1526) (SJ 14n3).

41. Schwartz, *Samuel Johnson* 73, 30.

42. A comparison to his measurements when he was on tour with the Thrales in 1774 suggests that Johnson is being deliberate here. See Schwartz on Johnson's "virtuoso" reports (ibid. 30).

43. Like Ian Duncan, I am interested here in Johnson's relation to "universal, abstract sign-systems" ("Pathos of Abstraction" 53). I argue that he formulates a different relation to the system of measurement, not through feeling, but through the presence of the physical body.

44. Schwartz, *Samuel Johnson* 7.

45. R. D. Connor, *Weights and Measures of Scotland* (Aberdeen: Her Majesty's Stationery Office, 1987) 245.

46. Ibid. 245.

47. I am suggesting that we should attempt to work around the extent to which, for those with a deep knowledge of Johnson's biography, this description has *become* Johnson. As Pat Rogers notes, "This was the first detailed physical picture he had supplied of his subject; indeed, it was the first time in a prominent and serious work of biography that the characteristic Johnsonian image which we all carry with us had been set out with such particularity" (*Johnson and Boswell* 98).

48. Helen Deutsch and Felicity Nussbaum, introduction to *"Defects": Engendering the Modern Body*, ed. Helen Deutsch and Felicity Nussbaum (Ann Arbor: University of Michigan Press, 2000) 2.

49. For examples from letters to Thrale, see Samuel Johnson, *The Letters of Samuel Johnson*, vol. 2, ed. Bruce Redford (Princeton, NJ: Princeton University Press, 1992) 62, 75, 98. Johnson customarily writes to Boswell in such a fashion, as when he records a cough having kept him from communicating about the publication of the *Journey* (123).

50. Rogers, *Johnson and Boswell* 23.

51. Deutsch and Nussbaum, introduction 1–2.

52. Ibid. 2.

53. Lennard J. Davis, "Dr. Johnson, Amelia, and the Discourse of Disability," in Deutsch and Nussbaum, eds., *"Defects"* 60.

54. Ibid. 61.

55. Ibid. 60–61.

56. See Helen Deutsch, "The Author as Monster: The Case of Dr. Johnson," in Deutsch and Nussbaum, eds., *"Defects"* 177–212. And Helen Deutsch, *Loving Dr. Johnson* (Chicago: University Chicago Press, 2005) esp. chap. 2.

57. William Hay, *Deformity: An Essay* (London, 1754) 7.

58. Ibid. 7.

59. Ibid. 20.

60. Ibid. 12.

61. Ibid. 12–13.

62. Davis, "Dr. Johnson" 56.

63. Garland-Thomson, ed., *Freakery: Cultural Spectacles of the Extraordinary Body*, qtd. in Deutsch and Nussbaum, introduction 7.

64. See Jerome Jeffrey Cohen, "Monster Culture (Seven Theses)," *Monster Theory: Reading Culture* (Minneapolis: University of Minnesota Press, 1996) 3–25.

65. Turnbull, "'Generous Attachment'" 227.

66. Andrew Curran and Patrick Graille, "The Faces of Eighteenth-Century Monstrosity," *Eighteenth-Century Life* 21.2 (1997): 3.

67. On literature's relation to participant observation see, for instance, James Buzard *Disorienting Fiction: The Authoethnographic Work of Nineteenth-Century British Novels* (Princeton, NJ: Princeton University Press, 2005).

68. Ryle, "Knowing How" 3.

69. Lennard J. Davis, *Enforcing Normalcy: Disability, Deafness, and the Body* (New York: Verso, 1995) 51.

70. Ibid. 53.

71. Ibid. 55.

72. Ibid. 59.

73. See, especially, Deidre Lynch, "'Beating the Track of the Alphabet': Samuel Johnson, Tourism, and the ABCs of Modern Authority," *ELH* 57. 2 (1990): 392–94 and Trumpener, *Bardic Nationalism*, both of whom argue for Johnson's focus on written language.

74. From his brief account of the "son of a constable of Spain," it appears that Johnson may have read John Bulwer's *Chirologia: Or, the Naturall Language of the Hand* (London, 1644). See Jason S. Farr's account of Bulwer in *Novel Bodies: Disability and Sexuality in Eighteenth-Century British Literature* (Lewisburg, PA: Bucknell University Press, 2019) chap. 1.

75. Davis, *Enforcing Normalcy* 71. Johnson is thus closer to Davis's own conclusion about what sign language could do: bridge the gap between oral and written language that plagued the Enlightenment.

76. Ibid. chap. 3.

77. Here, and elsewhere in this chapter, my account of Johnson is at some distance from Ian Duncan's account of Johnson as invested in a "vision that surmounts the limits of the body" ("Pathos of Abstraction" 53).

78. See Farr's account of Duncan Campbell (*Novel Bodies* chap. 1).

79. Trumpener, *Bardic Nationalism* 97.

80. Ibid. 100.

81. Farr, *Novel Bodies* 45.

82. Martin, preface 309.

83. Ibid. 300.

Chapter 6

1. Bruno Latour, *On the Modern Cult of the Factish Gods*, trans. Catherine Porter and Heather MacLean (Durham, NC: Duke University Press, 2010) esp. 1–6.

2. Ibid. 3; emphasis in original.

3. Other critics, too, have shown the ways Equiano writes back to the Enlightenment. For an account of Equiano's "shuttling ironic humor" and his "sly, subversive . . . narrative technique," see Chinosole, "Tryin' to Get over: Narrative Posture in Equiano's Autobiography," *The Art of Slave Narrative: Original Essays in Criticism and Theory*, ed. John Sekora and Darwin T. Turner (Macomb: Western Illinois University, 1982) 48, 47. For direct responses to Enlightenment philosophy, see Mark Stein, "Who's Afraid of Cannibals?

Some Uses of the Cannibalism Trope in Equiano's *Interesting Narrative*," *Discourses of Slavery and Abolition: Britain and its Colonies, 1760–1838*, ed. Brycchan Carey, Markman Ellis, and Sara Salih (New York: Palgrave Macmillan, 2004) 96–107; and Kerry Sinanan, who shows Equiano's employment of Rousseau's concepts and his text as a response to "the doubts that a supposedly civilized west was having about itself" ("The Slave Narrative and the Literature of Abolition," *The Cambridge Companion to the African American Slave Narrative*, ed. Audrey Fisch [Cambridge: Cambridge University Press, 2007] 65).

4. One of the most in-depth accounts of Equiano as a philosopher is April Langley's in *The Black Aesthetic Unbound*, where she follows the lead of Paul Edwards and Rosalind Shaw (who write on Equiano's "chi") in arguing for the presence of an African epistemology in the *The Interesting Narrative*. As in Edwards's and Shaw's earlier example, the argument is hard to support, given that it is very difficult to pin down abstract philosophical conceptions (e.g., "rupture and disjunction"; "resistance to totalizing structures"; "duality") as particularly African. That said, Langley's reads *The Interesting Narrative* to point outside the strictures of conventional European formulations, and ultimately concludes that Equiano does so as a way of "loosening fixed conditions," which is consonant with my account here. April C. E. Langley, *The Black Aesthetic Unbound: Theorizing the Dilemma of Eighteenth-Century African American Literature* (Columbus: Ohio State University Press, 2007) 111, 103, 134. Paul Edwards and Rosalind Shaw, "The Invisible Chi in Equiano's 'Interesting Narrative,'" *Journal of Religion in Africa* 19.2 (1989): 146–56.

5. Richard Sennett, *The Craftsman* (New Haven, CT: Yale University Press, 2008) 8.

6. Geraldine Murphy, "Olaudah Equiano, Accidental Tourist," *Eighteenth-Century Studies* 27.4 (1994) 561.

7. Ottobah Cugoano, *Thoughts and Sentiments on the Evil and Wicked Traffic of the Slavery and Commerce of the Human Species* (London, 1787).

8. Vincent Carretta, *Equiano, the African: Biography of a Self-Made Man* (London: Penguin, 2007). Most critics have assumed that Equiano's definition of the subject ends here. See, for instance, Werner Sollors, *African American Writing: A Literary Approach* (Philadelphia: Temple University Press, 2016) 22.

9. In this turn toward subjectivity as doing in the world, Equiano's account resembles recent attempts to rethink Black subjectivity phenomenologically. See, for example, Michelle M. Wright, *Physics of Blackness: Beyond the Middle Passage Epistemology* (Minneapolis: University of Minnesota Press, 2015); and Stephen Best, *None Like Us: Blackness, Belonging, Aesthetic Life* (Durham, NC: Duke University Press, 2018).

10. Roxann Wheeler, *The Complexion of Race: Categories of Difference in Eighteenth-Century British Culture* (Philadelphia: University of Pennsylvania Press, 2010) 262.

11. George E. Boulukos, "Olaudah Equiano and the Eighteenth-Century Debate on Africa," *Eighteenth-Century Studies* 40.2 (2007): 245.

12. Equiano very precisely sets Essaka in Smith's third stage of agriculture but on the verge of the fourth stage of commerce, when he notes that the society is agricultural with "few manufactures," which "make no part of our commerce" (Olaudah Equiano, *The Interesting Narrative* [ed. Brycchan Carey [Oxford: Oxford University Press, 2018]); hereafter cited in the text as *IN*. Before exchanging with other societies, Adam Smith remarks, "They would exchange with one an other what they produced more than was necessary for

their support" (*Lectures on Jurisprudence*, 1762–63; ed. R. L. Meek, D. D. Raphael, and P. G. Stein [Indianapolis: Liberty Fund, 1982] 15).

13. On the former see, e.g., Thomas Clarkson's explicit references to "stage of society" in discussing Africans (*An Essay on the Slavery and Commerce of the Human Species, Particularly the African*, 2nd ed. [London, 1788] 40). On the latter see, for example, the end of James Tobin's *Cursory Remarks Upon the Reverend Mr. Ramsay's Essay on the Treatment and Conversion of African Slaves in the Sugar Colonies* (London, 1785); and Gordon Turnbull, *An Apology for Negro Slavery: Or the West-India Planters Vindicated from the Charge of Inhumanity* (London, 1786).

14. William Pietz, "Bosman's Guinea: The Intercultural Roots of an Enlightenment Discourse," *Comparative Civilizations Review* 9.9 (1982): 2.

15. Ibid. 4.

16. William Pietz, "The Problem of the Fetish, I," *RES: Anthropology and Aesthetics* 9 (1985): 10, 10, 7, 10.

17. Pietz, "Bosman's Guinea" 3.

18. Pietz, "Problem of the Fetish" 9.

19. Pietz, "Bosman's Guinea" 3, 2.

20. Ibid. 4.

21. Ibid. 5.

22. Ibid. 5–6.

23. Endeavoring to pinpoint how it is, exactly, that one could understand an object to cause effects, to work as "magic," Hume would maintain, as does Bosman, that "superstition . . . rests on principles of the anthropomorphization of natural things and the subjection of human purpose to determination by the accidental events of nature" (ibid. 14).

24. Pietz, "Problem of the Fetish" 10.

25. Pietz, "Bosman's Guinea" 8–9.

26. William Bosman, *A New and Accurate Description of the Coast of Guinea, Divided into the Gold, the Slave, and the Ivory Coasts* (London, 1705) 157.

27. Pierre Bayle, *Oeuvres Diverses*, vol. 3 (Hildesheim: Georg Olm, 1966) 970–72; Pietz, "Bosman's Guinea" 6.

28. Bosman, *New and Accurate Description* 150.

29. As Sarah Tindal Kareem has observed, the aesthetic category of "wonder" proves powerful during this period in part because it opens up to a spectator not defined by social status. Wonder, she says, promotes "an indeterminate relationship between such responses and the types of people who might engage in them" (*Eighteenth-Century Fiction and the Reinvention of Wonder* [Oxford: Oxford University Press, 2014] 69).

30. See, for example, Walter Johnson, "Time and Revolution in African America: Temporality and the History of Atlantic Slavery," *Rethinking American History in a Global Age*, ed. Thomas Bender (Berkeley: University of California Press, 2002) 148. I thus find Equiano attentive to the kind of bifurcation that Johnson notices from his later historical perspective in the "contention of these temporal narratives," African and European, in the *Interesting Narrative*'s account of slavery (149).

31. See Ronald Paulson, *The Beautiful, Novel, and Strange: Aesthetics and Heterodoxy* (Baltimore: Johns Hopkins Press, 1996).

32. Henry Louis Gates Jr., *The Signifying Monkey: A Theory of African-American Literary Criticism* (Oxford: Oxford University Press, 2014) 169.

33. Lynn Festa, *Sentimental Figures of Empire in Eighteenth-Century Britain and France* (Baltimore: Johns Hopkins Press, 2006) 136.

34. Ibid. 137.

35. Bosman, *Coast of Guinea* 150.

36. Pietz, "Problem of the Fetish" 10.

37. Ibid.

38. Bosman, *Coast of Guinea* 150.

39. Ibid. 151.

40. "Watch, N.," *Oxford English Dictionary* (Oxford: Oxford University Press, June 2024) https://doi.org/10.1093/OED/1093718543.

41. This is a term that Werner Sollors uses to describe Equiano's comparisons of "African and English" (*African American Writing* 29–30).

42. Houston A. Baker Jr., *Blues, Ideology, and Afro-American Literature: A Vernacular Theory* (Chicago: University of Chicago Press, 1984) 37; emphasis in original.

43. Pietz, "Bosman's Guinea" 4.

44. Festa, *Sentimental Figures* 132. Festa and Jonathan Lamb pull this logic into their analyses of Equiano because they both begin with the "it-narrative," whose logic is that of the commodity (*The Things Things Say* [Princeton, NJ: Princeton University Press, 2011] chap. 10).

45. Michael Taussig, *Mimesis and Alterity: A Particular History of the Senses* (London: Routledge, 1993) 2.

46. Latour, *Modern Cult* 4.

47. It has become commonplace to refer to Equiano and his text in terms related to much later anthropology: "ethnologist" Ian Frederick Finseth, "In Essaka Once: Time and History in Olaudah Equiano's Autobiography," *Arizona Quarterly: A Journal of American Literature, Culture, and Theory* 58.1 (2002) 21; "ethnographic" Elizabeth A. Bohls, *Slavery and the Politics of Place: Representing the Colonial Caribbean, 1770–1833* (Cambridge: Cambridge University Press, 2014) 127; "ethnographer" Murphy 564; "cross-cultural readings," Stein, "Who's Afraid?" 101.

48. Finseth, "In Essaka Once" 1, 18, 18. This assumption also grounds Emily Donaldson Field's analysis of the way in which Equiano replaces Africans with Native Americans in the lowest stage of society. Field, "'Excepting Himself': Olaudah Equiano, Native Americans, and the Civilizing Mission," *MELUS: Multi-Ethnic Literature of the U.S.* 34.4 (2009): 15–38. See Adam Potkay for the argument about progress and Christianity ("Olaudah Equiano and the Art of Spiritual Autobiography," *Eighteenth-Century Studies* 27.4 [1994]: 677–92).

49. Gates, *Signifying Monkey* 171.

50. See, for example, Elizabeth A. Bohls, *Slavery and the Politics of Place: Representing the Colonial Caribbean, 1770–1833* (Cambridge: Cambridge University Press, 2014) 135.

51. Gates, *Signifying Monkey* 171.

52. See Finseth on parallax ("In Essaka Once" 10–11).

53. We might also note that this disturbs the temporal, progressivist assumptions of the autobiographical form in which Equiano writes. He is here closer to Wright's account of the "now" (*Physics of Blackness* 4).

54. Simon Gikandi, *Slavery and the Culture of Taste* (Princeton, NJ: Princeton University Press, 2011) 176.

55. Ibid. 176, 218.

56. Ibid. 218. Here Gikandi follows Hortense J. Spillers's account of the Middle Passage ("Mama's Baby, Papa's Maybe: An American Grammar Book," *Diacritics* 17.2 (1987): 64–81). See also Helen Thomas, *Romanticism and Slave Narratives: Transatlantic Testimonies* (Cambridge: Cambridge University Press, 2000) 229–30.

57. Orlando Patterson, *Slavery and Social Death: A Comparative Study* (Cambridge, MA: Harvard University Press, 1982) 5.

58. Langley, *Black Aesthetic* 113. See also Pier M. Larson on Equiano's combination of "familiarity and strangeness" ("Horrid Journeying: Narratives of Enslavement and the Global African Diaspora," *Journal of World History* 19.4 [2008]: 454).

59. Pietz, "Bosman's Guinea" 2–3.

60. Here I depart from the most extensive critical argument on the fetish and Equiano in Srinivas Aravamudan's *Tropicopolitans*. Aravamudan demonstrates the extent to which Equiano's understanding of the book as fetish calls into question the accounts of a literacy-based subjectivity earlier critics like Gates imagined for him (*Tropicopolitans: Colonialism and Agency, 1688–1804* [Durham, NC: Duke University Press, 1999] 269). In turning to the fetish as a matter not of reading but of bodily perception we find an additional attention to the material body. Monique Allewaert leaves Equiano with the book-as-fetish and but her other readings of Enlightenment writers, who engaged the African fetish in terms of the "materialized black body," are closer to my reading here (*Ariel's Ecology: Plantations, Personhood, and Colonialism in the American Tropics* [Minneapolis: University of Minnesota Press, 2013] 127).

61. Bohls also remarks on an "incongruous shift" later in the book, at the beginning of chapter 5, when "Equiano's narrative suddenly swerves from the perspective of the enslaved African majority to that of a privileged European" (*Slavery and the Politics of Place* 141). Rather than resting with the "ambiguity" Bohls concludes for it, we could read that passage—chapter 5 begins, after all, with an account of the slave torture Equiano will not relate, and thus a heightened awareness of the Black body—through the logic of reembodiment that I describe here.

62. Best, *None Like Us* 125.

63. Festa, *Sentimental Figures* 138.

64. Latour, *Modern Cult* 11.

65. Brycchan Carey, *British Abolitionism and the Rhetoric of Sensibility: Writing, Sentiment and Slavery, 1760–1807* (New York: Palgrave MacMillan, 2005) 137.

66. Ibid. 140, 137. For a more typical case of sentimental rhetorical appeals, see John Wesley, *Thoughts upon Slavery* (London, 1774).

67. Olaudah Equiano, "To Mr. Gordon Turnbull, Author of an 'Apology for negro slavery" (1788; *The Interesting Narrative and Other Writings*, ed. Vincent Carretta [New York: Penguin, 2003] 334).

68. Ramesh Mallipeddi, *Spectacular Suffering: Witnessing Slavery in the Eighteenth-Century British Atlantic* (Charlottesville: University of Virginia Press, 2016) 2.

69. Ibid. 3; for Mallipeddi's critique of Hartman see 13–14.

70. Saidiya V. Hartman, *Scenes of Subjection: Terror, Slavery, and Self-Making in Nineteenth-Century America* (New York: Oxford University Press, 1997) 3, 4.

71. Ibid. 18.

72. Ibid.

73. Ibid. 19.

74. Ibid.

75. See Festa's account of sympathy as appropriation (*Sentimental Figures* esp. chap. 4).

76. Clarkson, *Essay on the Slavery* 87.

77. Ibid. 88.

78. Ibid.

79. Marcus Rediker, *The Slave Ship: A Human History* (New York: Viking Penguin, 2007) 109.

80. William Falconer, *An Universal Dictionary of the Marine* (London, 1780).

81. Roland Barthes, "The Reality Effect," in *The Rustle of Language*, trans. Richard Howard (New York: Hill and Wang, 1986) 141.

82. Ibid. 141, 143.

83. Elaine Scarry, *The Body in Pain: The Making and Unmaking of the World* (New York: Oxford University Press, 1985) 41. For an eighteenth-century analog, see Clarkson's account of the weapons of slavery: "any thing that passion could seize, and convert into an instrument of punishment, has been used" (*Essay on the Slavery* 110).

84. Rediker, *Slave Ship* 17; Marcus Rediker, *Between the Devil and the Deep Blue Sea: Merchant Seamen, Pirates and the Anglo-American Maritime World, 1700–1750* (Cambridge: Cambridge University Press, 1987) 155.

85. W. Jeffrey Bolster, *Black Jacks: African American Seamen In the Age of Sail* (Cambridge, MA: Harvard University Press, 1997) 2.

86. The windlass, I am suggesting, offers us on the level of textual detail the larger claim that Peter Linebaugh and Marcus Rediker, and after them, Matthew D. Brown, describe as the liberating potential of maritime life (*The Many-Headed Hydra: Sailors, Slaves, Commoners and the Hidden History of the Revolutionary Atlantic* [Boston: Beacon, 2000]). Brown argues that in the *Interesting Narrative* in particular, "it is through the ship and the sea that Equiano is able to rewrite his structural relation to slavery" ("Olaudah Equiano and the Sailor's Telegraph: 'The *Interesting Narrative*' and the Source of Black Abolitionism," *Callaloo* 36.1 (2013): 193.

87. William Falconer, "Windlass," in *A Universal Dictionary of the Marine* (London, 1780) n.p.

88. Rediker, *Between the Devil* 90.

89. Ibid. 160, 173.

90. Margaret Cohen, *The Novel and the Sea* (Princeton, NJ: Princeton University Press, 2010) passim.

91. Ibid. 17.

92. Ibid. 55–58.

93. Ibid. 20–21.

94. Ibid. 25.

95. Ibid. 31, 32.

96. Although Vincent Carretta does not bring in maritime craft, he similarly moves away from the text's own insistence on its commitment to spiritual autobiography, giving us Equiano's experience in strictly psychological terms (introduction to *The Interesting Narrative of the Life of Olaudah Equiano, and Other Writings*, ed. Vincent Carretta [New York: Penguin, 2003] ix–xxx).

97. Debbie Lee and Louis Kirk McAuley argue that we are asked to read in the plate of the wreck "an idea of the end of the transatlantic slave trade." If so, then my argument suggests that Equiano envisions a different "end" than the "ideal African consumer" ("Romantic Recycling: The Global Economy and Secondhand Language in Equiano's *Interesting Narrative* and the Letters of the Sierra Leone Settlers," *Global Romanticism: Origins, Orientations, and Engagements, 1760–1820*, ed. Evan Gottlieb [Lewisburg, PA: Bucknell University Press, 2015] 176, 186).

98. George Boulukos argues that "this moment can be taken as one of the strongest examples of Equiano's persistent implications that colonial whites themselves understand white supremacy as a matter of merely material advantage by force, rather than a reflection of essential differences between black and white" (*The Grateful Slave: The Emergence of Race in Eighteenth-Century British and American Culture* [Cambridge: Cambridge University Press, 2008] 197).

99. "Equiano appears to offer the transformation of his own attitude toward the varieties of eighteenth-century slavery as a model for the moral progress of his readers as individuals and of the society he now shares with them" (Carretta, "Introduction" xxii). Equiano's biography is complicated on this score. He had a stint as an overseer on a slave plantation in South American in 1775 and by 1776 still did not oppose slavery.

100. Carretta, *Equiano*.

101. Ian Baucom, *Specters of the Atlantic: Finance Capital, Slavery, and the Philosophy of History* (Durham, NC: Duke University Press, 2005).

102. Ibid. 4, 11.

103. Ibid. 11–14.

104. Ibid. 42–43.

105. See the theologian Willie James Jennings's account of Equiano's realization of the limitations of Christianity in his eighteenth-century context, which can afford him "belonging to God alone" rather than the "love between peoples" and larger community for which Equiano might have wished (*The Christian Imagination: Theology and the Origins of Race* [New Haven, CT: Yale University Press, 2010] 202, 201). Following Jennings's reading, we can see Equiano move Christianity toward another kind of communal solution in this passage.

106. Rediker, *Between the Devil* 106.

107. Ibid. 110.

108. Rediker and Linebaugh, *Many-Headed Hydra* 191, 193. On slave rebellions contemporary with Equiano's, see Michael Craton, *Testing the Chains: Resistance to Slavery in the British West Indies* (Ithaca, NJ: Cornell University Press, 1982). We might, then, pair

the passage with Beelzebub's speech from *Paradise Lost*, which Equiano borrows for the end of chapter 5. Vincent Carretta notes about this passage that it "leaves open the option of violent resistance" ("Equiano's Paradise Lost: The Limits of Allusion in Chapter Five of *The Interesting Narrative*," in *Imagining Transatlantic Slavery*, ed. Cora Kaplan and John Oldfield [New York: Palgrave Macmillan, 2010] 93).

109. Peter Linebaugh, *The London Hanged: Crime and Civil Society in the Eighteenth Century* (London: Verso, 2006) 415.

110. John Bugg, "The Other Interesting Narrative: Olaudah Equiano's Public Book Tour," *PMLA* 121.5 (2006): 1427. Equiano's political strategy seems to resemble his literary one, sneaking in a more radical message than the one he appeared to deliver, as in the "supplementary abolitionist pamphlets he quietly distributed," including Fox's tract on the sugar boycott, known for its argument that consuming sugar was consuming African "flesh" (ibid. 1431).

111. We might compare this positive claim to the other injuries in the text which, as Jonathan Elmer claims, seem to disappear. Elmer treats the "inaccessibility" of Equiano's "wounds and scars" through trauma theory (*On Lingering and Being Last: Race and Sovereignty in the New World* [New York: Fordham University Press, 2008] 73).

112. Hartman, *Scenes of Subjection* 54.

113. Ibid. 69.

114. Sennett, *Craftsman* 6.

115. Ibid.

116. Ibid. 6–7.

Coda

1. Rita Felski, *The Limits of Critique* (Chicago: University of Chicago Press, 2015) 12, 173. Felski draws on and responds to Bruno Latour's influential essay, "Why Has Critique Run Out of Steam? From Matters of Fact to Matters of Concern," *Critical Inquiry* 30 (2004): 225–48. In her attention to "assembling," one can hear an echo of Latour's book title, *Reassembling the Social: An Introduction to Actor-Network-Theory* (Oxford: Oxford University Press, 2007).

2. Patrick Jagoda, "Critique and Critical Making," *PMLA* 132.2 (2017): 356–63.

3. David Staley, "On the 'Maker Turn' in the Humanities" *Making Things and Drawing Boundaries*, ed. Jentery Sayers (Minneapolis: University of Minnesota Press, 2018) 33.

4. Jonathan Kramnick, *Criticism and Truth: On Method in Literary Studies* (Chicago: University of Chicago Press, 2023) 74.

5. One doesn't have to reach far for a comparison of Kramnick to the natural philosophers of this book. Kramnick is a specialist in eighteenth-century literature, and as part of his book-length argument asks us to sign on to Jonathan Richardson's 1725 account of the painter (restating the terms of earlier natural philosophers like Hooke), who "must be a mechanic, his hand and eye . . . as expert as his head is clear and lively" (94).

6. Jagoda, "Critique and Critical Making," 358.

7. Kramnick describes this "ethos" as involving "everyday science and ordinary expertise" (4).

8. Here I have in mind Felski's earlier work in *Uses of Literature* (New York: Wiley-Blackwell, 2008).

9. For a fuller account, see the introduction.

10. I speculated on the relationship between Collier, craft and academic work in "What Kind of Future Are We Making?" *The Rambling* 9 (August 7, 2020) https://the-rambling.com/2020/08/07/issue9-mack/ (accessed September 10, 2024).

Index

Adamson, Glenn, 253n3, 258n90
Addison, Joseph: "The Pleasures of the Imagination" essays, 99, 101, 170–74, 255n30; on the georgic, 67–68, 74; *The Spectator*, 1–3, 24, 25, 233n1, 233n2, 234n3, 234n4
aesthetic philosophy, 9, 169–70, 201; Addison, Joseph, 99, 101, 170–71, 255n30; Berkeley, George, 99–101; Burke, Edmund, 201; Hogarth, William, 95–127; Hutcheson, William, 99–101; Kant, Immanuel, 101, 255n3; Rancière, Jacques, 71; Shaftesbury, Anthony Ashley Cooper, 3rd Earl of, 99; Reynolds, Joshua, 263n71
Ahmed, Sara, 167
Akhimie, Patricia, 80
Alberti, Leon Battista, 119, 240n79, 257n83
Alff, David, 246n1
Allewaert, Monique, 272n60
Alpers, Paul, 172
Alpers, Svetlana, 20, 46, 121, 240n70, 244n45, 258n89
Andrews, Corey E., 70

anthropology, 16, 24–25, 148, 152, 155, 204, 205, 271n47; and colonialism, 259n5, 272n60; and the contact zone, 128, 138, 152; ethnography, 128, 147, 200, 205, 259n1, 261n44, 271n47; history of, 2, 23–26, 128, 197; theories of, 25, 125–27, 142–43, 155, 228; and the participant-observer, 2, 126, 142–43, 187, 261n44, 268n67. *See also* fetishism
antipastoral, 72–74, 77, 89
antiquarianism, 175–77
antislavery movement: Benezet, Anthony and, 196; Equiano, Olaudah and, 29, 196–97, 202, 204, 211, 212, 214, 216, 223, 226, 228, 273n86, 275n110; and sentimentality, 212, 272n66; the sugar boycott, 275n110; Wesley, John, and 272n66; and the *Zong* massacre, 224
Aram, Peter, 65
Arendt, Hannah, 81, 228
Aristotle, 2–3, 8, 10, 15, 97, 109, 110, 236n26
art: as applied or technical knowledge, 6, 63–64, 95, 120–21; split between beaux arts (fine arts) and useful arts (crafts) 7, 95, 254n12

Index

artisan: 97, 107; aesthetic spectator as, 71, 112; in Bacon, Francis, 2–3, 8–11, 107; in capitalism, 5, 13; habit as, 55; and language, 17; mariner as, 147–48, 195, 220–23; painter as, 98; poet as, 65–68; as speculative, 1–2, 3, 233n2; in Sprat, Thomas, 2–3, 8–11, 107; thief as, 32, 43, 49; woman as, 81–94; writer as, 107
Astley, Thomas, 198
astronomy: Defoe, Daniel and, 32, 35; and navigation, 132, 147–48; and the Royal Society, 96; the Transit of Venus, 132, 135, 260n13
Aravamudan, Srinivas, 272n60

Bacon, Francis, 2–3, 6, 8–11, 13, 18–20, 33–39, 55, 63, 95–96, 98, 107, 160
Backscheider, Paula R., 242n10
Baker, Henry, 96
Baker, Houston, 204
Bancks, John, 65
Banks, Joseph, 5, 16, 24, 26, 28, 128–57
Baridon, Michael, 254n20
Barney, Richard, 242nn 14, 16, 243n31
Barrell, John, 67, 69–70, 80, 97, 239n72, 248n24, 249n45, 250n60
Barthes, Roland, 45, 217
Bateson, Gregory, 25
Batt, Jennifer, 247n16
Baucom, Ian, 224–25
Bayle, Pierre, 193, 198–99
Beaglehole, J. C., 128, 132, 137, 260n13, 261n27
Behn, Aphra, 253n113
Benezet, Anthony, 195
Bennett, James A., 236n25
Benjamin, Walter, 15, 152, 204, 232, 239n70
Benton, Ted, 78
Berkeley, George, 100–101, 255n33
Bertucci, Paula, 234n6, 236n30
Bindman, David, 258n89
Blair, Hugh, 161, 164
Blewett, David, 42, 243n33, 245n54
Boerhaave, Herman, 160, 265n11

Boethius (Hector Boece), 175
Bolster, W. Jeffrey, 218
Bosman, William, 198–99, 202–3, 211
Boswell, James, 3, 24, 29, 128, 158–62, 165–67, 176–77, 181–93
Boulukos, George, 196, 246n65, 274n98
Bourdieu, Pierre, 25, 126, 155, 241n100, 243n31
Boyd, Bradford Q., 172–73
Boyle, Robert, 5, 12, 32, 47–48, 131, 160
Brathwaite, Richard, 242n13
Brewer, John, 5, 235n14
Briscoe, Robert, 119–21
Brown, Bill, 244n37
Brown, Laura, 92, 253n113
Brown, Matthew D., 273n86
Bugg, John, 227, 275n110
Bulwer, John, 268n74
Burke, Edmund, 171, 174, 178, 201, 266n39, 266n38
Butler, Judith, 243n31
Bysshe, Edward, 5, 64

Campbell, Duncan, 192, 268n78
Campbell, Mary Baine, 20
capitalism: agrarian, 69; and craft, 5, 13, 16, 232
Carey, Brycchan, 212–13
Carretta, Vincent, 195, 213, 224, 274n96, 274n99, 275n108
Carrithers, David, 23
Carter, Paul, 144, 156
Chalker, John, 67
Chambers, Ephraim, 6–7, 63, 95, 96
Charlevoix, Pierre-François-Xavier de, 162, 265n16
Chesterfield, Philip Stanhope, 2nd Earl of, 68
Christmas, William J., 64–65, 248n31, 249n45, 251n74
Chico, Tita, 12, 13, 51, 90, 91, 233n2, 239n73, 233n2
Cicero, Marcus Tullius (Tully), 42
Clark, Kenneth, 120

Clarkson, Thomas, 215, 270n13, 273n83
Clegg, Arthur, 236n25
Clewis, Robert R., 100, 255n31
cognitive ecologies, 15, 16, 238n65
cognitive science, 15, 16
Cohen, Jerome Jeffrey, 267n64
Cohen, Margaret, 14, 137, 220–21, 225, 238n63, 239n74, 242n19, 244n44
Cohen, Ralph, 72, 250n58
Collier, Mary, 3, 16–17, 27, 63, 64, 66–67, 81–94, 232
Collingwood, Luke, 224
colonialism, 6, 29, 146–48, 162–69. *See also* imperialism
Connor, R. D., 179
Cook, James (Captain), 23, 24, 28, 128–29, 131–32, 135, 147–48, 160, 221
Coppola, Al, 12–13, 51, 239n73
Cowley, Abraham, 18–19, 22, 46, 171
craft. *See* handicraft
craft knowledge (European), 3, 8, 103, 107; and aesthetics, 95–127, 170, 174; on the body, 144–47, 203; and crime, 43, 52; and domestic labor, 81; in education, 34, 55; and gender, 85; and habit and repetition, 55, 77, 94, 112; and husbandry, 63–94; 164–65; and handwriting, 38, 79, 108, 135, 154–55; and the humanities (contemporary), 230–32; and indigenous practices, 22, 146, 147, 165; beyond the individual, 22, 62, 141–42, 149, 155, 177, 188, 219, 226; and mastery, 88, 93, 145, 195, 212; and navigation, 147–48, 195, 220–23; and poetry, 64, 68, 76–77; and the printed book, 107; and slavery, 80, 212; and time, 77–78; and trade, 37, 41. *See* also embodied knowledge
Craton, Michael, 275n108
crime, 32, 47–51
Crombie, A. C., 236n25
Cugoano, Ottobah, 195
Culler, Dwight, 64, 247n6
Curran, Andrew, and Patrick Graille, 186

Dalrymple, Alexander, 132
Dampier, William, 23, 131
Daston, Lorraine, and Peter Galison, 139, 141, 152, 263n66
Davidson, Jenny, 241n4, 242n16, 243n32
Davis, Lennard, 183, 184, 189–91, 268n75
Dear, Peter, 21, 46–47, 51
de Brosses, Charles, 198
De Bruyn, Frans, 235n20, 246n1
de Certeau, Michel, 25, 126, 228, 241n101
Defoe, Daniel: *Captain Singleton*, 53; *Colonel Jack*, 26–27, 31–62; *The Complete English Gentleman*, 26, 33, 42, 51; *The Complete English Tradesman*, 26, 32–39; *Moll Flanders* 45, 53, 245n56; *Robinson Crusoe* 6, 20–22, 246
de l'Épée, Charles-Michel, 189–90
Deligny, Fernand, 126, 258n91
design, 97, 112, 254n15
Deutsch, Helen, 183
Deutsch, Helen, and Felicity Nussbaum, 182–83, 185
Dickie, George, 99
Diderot, Denis, 189
disability, 88, 182–85, 189–93, 268n74
Dodsley, Robert, 65, 76, 246n1, 250n54
Dolphin, the, 135, 136, 137
Donald, Diana, 99
Dormer, Peter, 20, 239n78
Dryden, John, 6, 65–68, 69, 75, 247n21
Duck, Stephen, 3, 27, 63–80
Duncan, Ian, 265n9, 267n43, 268n77

Eamon, William, 13, 134, 235n17, 261n20
Eckstein, Lars, and Anja Schwartz, 148, 262n61
education: Baconian influence on, 32–33; and craft, 55; Defoe, Daniel on, 32–41; Locke, John on, 33; and the novel, 42, 243n31
Edwards, Paul, and Rosalind Shaw, 269n4
Edwards, Philip, 6

Index

Embodied knowledge: in art (as craft), 107–23, 141; the criminal's, 54–62; and disability, 87–88, 182–85, 189–93; in drawing, 108, 148–55, 263n68, 263n66; and education, 35; as ethical, 15–16, 17, 239n72, 93–94, 129, 155–57, 193, 232; as failure, 182, 194; the laborer's, 66, 74–75, 81–94; and punishment, 62; and race, 225, 227; and sight, 71–74, 136, 140, 172, 177–78, 181, 192–93, 210–11; and slavery, 80, 195, 225; as social 3, 16, 39, 81, 87–88, 91, 159, 188, 218–19; the woman's, 81–94; in writing, 135. *See also* craft knowledge

Enlightenment, the (term), 6

episteme. *See* knowledge: theoretical

ethnography, 259n1, 200, 271n47

Equiano, Olaudah, 3, 29, 194–229

Evelyn, John, 9

experiment: and craft, 8, 137–42, 156–57; in Cowley, Abraham, 19; in Defoe, Daniel, 20–22, 47–53; in Duck, Stephen, 75

Fairer, David, 93, 248n24, 250n71, 252n111
Falconer, William, 215, 218
Faller, Lincoln, 53, 241n5, 245n54, 246n67, 276n8
Farr, Jason, 192, 268n74, 268n78
Felski, Rita, 230, 275n1
Ferguson, Adam, 16, 161
Festa, Lynn, 18, 202, 203, 204, 244n44, 271n44, 273n75
fetish, the, 194, 197–200, 202–3, 211–12
Fielding, Henry, 113
Fisher, Carl, 245n56
form, 103, 104–7, 171
Foucault, Michel, 138–40, 240n79
Fowler, Elizabeth, 173
Fox, Celina, 11–14, 122, 235n24, 239n77, 243nn5–6, 253n5
Fox, George, 275n110
Freeman, Lisa, 262n51
French Academy, the (Académie des Beaux-Arts), 97, 109

French Enlightenment, 234n6
Frizzle, John, 65

Gailhard, Jean, 242nn13–14
Galileo, 32, 35
Gamble, Ellis, 96
Garland Thompson, Rosemarie, 185
Gates, Henry Louis, Jr., 202, 205, 272n60
Geertz, Clifford, 25
Gell, Alfred, 146
georgic poetry, 81, 246n1, 247n20, 248n25; Addison, Joseph and, 67, 68; and artisanal knowledge, 24, 27, 55, 63, 64, 66–7, 88; and care, 93–94; and decorum, 68, 74; Dryden's translation, 65–69; the eco-georgic, 252n111; and embodiment, 68, 85, 89; and gender, 81–94; and intellectual labor, 27, 65–6, 68, 159; by laboring-class poets, 63–94; and natural philosophy, 67–78, 246n1; and the pastoral, 248n24; Virgil's, 27, 67–69, 93, 246n1. *See also* Collier, Mary; Duck, Stephen; Virgil
Gibbon, Edward, 198
Gibson, J. J., 118–21
Gikandi, Simon, 206, 272n56
Gilligan, Carol, 93–94, 253n117
Glykon (Glikon), 97, 104–7, 110, 123
Gombrich, E. H., 117–21
Goodman, Kevis, 67, 70, 77, 249n49
Goodridge, John, 71, 75, 82, 249n38, 252n98
Graham, George, 179–80
Gravelot, Hubert François, 126–27
Greenhalgh, Peter, 253n2
Grotius, Hugo, 42
Guest, Harriet, 154, 263n71
Guyer, Paul, 100, 171, 255nn31–33

Habermas, Jürgen, 7, 78
handicraft (term), 2, 8, 19. *See also* craft knowledge
Hanson, Craig Ashley, 258n91
Haraway, Donna, 234n3
Hart, Francis R., 264n9

Hartman, Sadiya, 214–16, 227–28
Harvey, William, 32
Hawkesworth, John, 132, 160, 221
Hayles, N. Katherine, 238n66
Hegel, Georg Wilhelm Friedrich, 78
Heidegger, Martin, 15, 239n70
Hickey, Alison, 266n35
Hobbes, Thomas, 4
Hodgen, Margaret T., 24
Hogarth, William, 3, 7, 26–28, 77; *Analysis of Beauty*, 95–127; *Gin Lane*, 113; *Gulielmus Hogarth*, 113, *114*, 115; *Harlot's Progress*, 113
Hooke, Robert, 9, 11–12, 20–21, 45
Holder, William, 190
Horace, 37, 64, 133, 242n20
Horkheimer, Max, and Theodor W. Adorno, 239n70, 364n79
how-to manuals: 63, 122, 235n17; Aber, Israel, *The Art of Manufacturing Saltpetre*, 5; Bysshe, Edward, *The Art of English Poetry*, 5, 64, 247n6; E. T., *The Art and Mystery of Vintners and Wine-Coopers*, 64; Gardiner, John Smallman, *The Art and the Pleasures of Hare-Hunting*, 64, 246n5; Keys, John, *The Practical Bee-Master*, 5; Kirby, Joshua, *Dr. Brook Taylor's Method of Perspective Made Easy*, 122, 258n91; Mitchell, John, *Ars Scribendi Sine Penna: Or, The Art of Taking Down Sermons, Trials, Speeches, &c. Verbatim Without Pen and Ink*, 64, 246n5. *See also* craft knowledge, print market
Hume, David, 4, 6, 23, 198, 270n23
Hunter, Matthew, 235n24; 254n17
Hunter, Michael, 11, 237n43
Hutcheson, Francis, 99–101, 107, 111, 255n27
Hutchins, Edward, 15, 238n65

imperialism, 5, 7, 16, 22, 29–30, 96, 129, 155–56, 186, 193, 226, 232, 260n18, 274n98

Industrial Revolution, 4–5, 13
Ingold, Tim, 116–17, 119, 125, 155, 256n55, 257n83, 259n100

Jackson, Michael, 25
Jagoda, Patrick, 230–31
James, Robert, 265n11
Jennings, Willie James, 274n105
Johnson, Samuel, 5, 6, 63, 158–93
Johnson, Walter, 270n30
Jonson, Ben, 250n51

Kames, Henry Home, Lord, 265n17
Kant, Immanuel, 101, 255n3
Kareem, Sarah Tindal, 270n29
Kashanipour, Jasamin, 152
Keegan, Bridget, 74, 79, 248n28
Kellett, J. R., 235n14
Kepler, Johannes, 32
Kirby, Joshua, 122, 258n91
knowledge, practical (*techne*). See Aristotle; craft knowledge; embodied knowledge
knowledge, theoretical (*episteme*), 8, 15, 25, 97, 126, 187–88. *See also* philosophy and mathematics
Knox, John, 175
Kramnick, Jonathan, 7, 14–15, 90–91, 118, 230–32, 238n66, 239n72, 275nn5
Kristeva, Julia, 90

Labat, Jean-Baptiste, 198
labor, intellectual, 11, 27, 64–66, 68, 78, 159, 247n20, 251n89
labor, physical, 276, 66–68, 77–78, 81–82, 85, 89, 91
laboring-class poetry, 64–65, 81–82. *See also* Aram, Peter; Bancks, John; Collier, Mary; Dodsley, Robert; Duck, Stephen; Frizzle, John; Tasterstal, Robert
Lafitau, Joseph-François, 162, 265n16
Lamb, Jonathan, 259n5, 262n53, 263n67, 271n44
Lamont, Claire, 265n17

Index

Landry, Donna, 86, 252n99
Langley, April, 207, 269n4
Larson, Pier M., 272n58
Lascelles, Mary, 264n9
Latour, Bruno, 132, 141, 155, 194, 198, 204, 211–12, 275n1
Leask, Nigel, 259n5
Lee, Debbie, and Louis Kirk McAuley, 274n97
Lembke, Janet, 93
Linebaugh, Peter, 226, 273n86. *See also* Rediker, Marcus, and Peter Linebaugh.
Linnaeus, Carl, 28, 138–39, 240n79
Locke, John, 27, 198; *Essay Concerning Human Understanding*, 34, 54–55, 100, 255n24; *Second Treatise of Government*, 164, 167; *Some Thoughts Concerning Education*, 6, 33–36, 39, 42–43, 53–55, 58–59, 124, 135, 242n14
Lomazzo, Gian Paolo, 99, 113
Long, Pamela O., 13, 236n25
Lorde, Audre, 238n68, 251n89
Lynch, Deidre, 268n73
Lynch, William T., 237n40
Lyttelton, George, 70

Macdonald, James, 165
Macpherson, James, 161
Mackie, Erin, 243n31
Makin, Bathsua, 242n13
Mallipeddi, Ramesh, 213, 214
Malone, Edward, 160
Manicas, Peter T., 234n7, 240n91
Martin, Martin, 160, 165, 166–67, 192
Martin, Peter, 264nn3–4
Marx, Karl, 13, 15, 69, 78, 238n69, 244n37, 251n78
mathematics, 106, 109–10
Mauss, Marcel, 241n100
McKentrick, Neil, 5, 235n14
McLynn, Frank, 244n35, 245n60
Merchant, Carolyn, 264n81
Merleau-Ponty, Maurice, 87–88
Miller, Philip, 158

Milton, John, 68, 115
"modest witness," the, 12, 21, 52, 53, 234n3; critiques of, 12–13, 14, 24, 51, 239n73; in the social world, 1–3, 53; and Haraway, Donna J., 234n3
Mokyr, Joel, 5–6, 13, 235n18
Monboddo, James Burnett, Lord, 158, 161, 182, 189, 264n2
Monk, Samuel, 243n31
Montague, Charles, 131
Montesquieu, Charles Louis Secondat, baron de, 4, 16, 49
Morton, Charles, 32–3, 35, 42, 242n8
Morton, James Douglas, Earl of, 134, 240n94
Moxon, Joseph, 14, 19, 20
Mun, Thomas, 243n25
Murphy, Geraldine, 195, 271n47

natural history, 28, 138–39, 209–10, 240n79. *See also* Linnaeus, Carl
natural philosophy: and education 32–33; and the georgic 63, 67; and the recipe 134–35, 137; and the social world 39, 49, 53, 57; and taxonomy 138–42; and travel writing 130–36, 160. *See also* Royal Society; Bacon, Francis; Sprat, Thomas
Netz, Reviel, 110, 257n64
Newton, Isaac, 4, 23, 32, 160
Noggle, James, 123
Novak, Maximillian E., 20, 45, 244n38

objectivity, 11, 25, 139, 175, 210, 260n12
O'Briain, Katarina, 247n21
O'Brian, Patrick, 261n24
O'Brien, Karen, 251n83
Ogée, Frédéric, 255n22
Oliver, Douglas L., 143
Ossian, 161, 164. *See also* Macpherson, James

Parkinson, Sydney, 138, 140, 141, 148–49, 259n4, 263n71

pastoral, 70–74, 77–79, 172–73, 179, 248n24, 248n31, 250n58. *See also* antipastoral; Pope, Alexander; Philips, Ambrose; Tickell, Thomas
Patterson, Orlando, 206
Paulson, Ronald, 109, 111–12, 122, 235n22, 254n11, 254n13, 255n27, 258n89
Peacham, Henry, 33, 55, 242n13
Pearl, Jason, 131, 260n12
Pellicer, Juan Christian, 246n1, 250n54
Pennant, Thomas, 138, 160–61, 175
Pérez-Ramos, Antonio, 236n31, 237n39
phenomenology, 238n67, 87–88, 118–20, 156, 177, 232
Philips, Ambrose, 73
Philips, John, 158
philosophy: and Equiano, Olaudah, 269n4; and handicraft 2; and Johnson, Samuel, 159, 160, 168, 188; Sprat, Thomas and Addison, Joseph on, 233n1; as a way to know, 2. *See also* knowledge, theoretical; Aristotle
Picciotto, Joanna, 11, 13
Pietz, William, 194, 197–99, 202, 208–9
Pliny, 18
Poovey, Mary, 39–41, 239n74, 243n25, 264n9
Pope, Alexander, 73, 92, 247n19, 250n60, 258n112
Plumb, J. H., 5, 235n14
Pratt, Mary Louise, 138–39
print market, 5, 13, 14, 63–64
Price, Richard, 13
Pufendorf, Samuel, 42
Pumfrey, Stephen, 13
punishment: Bloody Code, 59; in education, 58–59; public, 59, 61–62; and slavery, 61–62, 213–14; and society, 57
Pye, David, 20, 239n78

race: and customs, 196–97; and chattel slavery, 80; and embodiment, 223, 225–26, 227; and laborers in Britain, 80, 89, 90, 251n83; and sympathy 214
Radner, John B., 159, 264n7

Rancière, Jacques, 71, 72, 250n56
realism (literary): and action, 46–53, 56–57; and craft, 22, 62, 168, 171; and the georgic, 68; in the novel, 20–22, 31–32, 45–53, 62; and number, 224–25; and objects, 45, 241n4, 244n37, 216–17; and painting, 20, 45–46, 240n79; and the plain style, 20, 56–57, 62, 168; and society, 57, 62; and violence, 54–62, 217. *See also* representation
recipe, 21, 133–34, 137
Rediker, Marcus, 216, 217, 218, 220
Rediker, Marcus, and Peter Linebaugh, 226, 273n86
relativism, 101, 107
Rennie, Neil, 261n24, 261n43
representation: and action and experiment, 17, 20, 22, 46–53, 57; and the body, 7, 14, 18–19, 20, 40, 108, 110; and facticity, 39–40, 131; and imperialism, 131–32, 155–57; and painting and engraving, 18–19, 20, 107–23, 138–42; and performance, 142–48, 262n58; and plain language or style, 7, 17, 31–32, 37–38, 39, 56–57, 62, 129, 130–31, 137; and society, 57, 62. *See also* realism
Reynolds, Sir Joshua, 7, 97, 254n17, 258n89, 263n71
Richardson, Jonathan, 275n5
Richetti, John, 60, 242n10, 243n31, 245n54
Richardson, Samuel, 20, 45
Risatti, Howard, 103–4, 106, 254n12, 256n44
Robinson, Nicolas, 84
Rogers, Pat, 160, 264n4, 265n10, 265n17, 268n47
Rossi, Paulo, 236n25, 236n27
Royal Academy of Arts, the, 7, 95, 96–97
Royal Society, the, 3, 97, 98, 160; and education, 32; and experiment, 12, 19, 21, 47–53, 156–57; and history of trades, 9; and language, 17, 131, 190; and tradesmen, 11; and travel writing, 128, 131, 175
Ryle, Gilbert, 5, 122, 168, 188–89, 235n12

Salamon, Gayle, 88
Sandwich, John Montagu, 4th Earl of, 138
Sayers, Sean, 78, 238n69, 251n78
Scarry, Elaine, 217
Schmidgen, Wolfram, 241n4, 244n37
Schwartz, Richard, 160, 175–76, 177, 179, 265n11, 266n28, 266n42
science. *See* natural philosophy
Scottish Enlightenment, 4, 161–62, 163–64, 197, 265n17, 269n12, 271n48
Sennett, Richard, 16, 94, 156, 195, 228, 236n27
sentiment, 212–13
Shadwell, Thomas, 12–13, 233n2
Shaftesbury, Anthony Ashley Cooper, 3rd Earl of, 99, 101, 255n27
Shapin, Steven, 12, 13
Shapin, Steven, and Simon Schaffer, 12–13, 20–21, 234n3, 240n79
Sharp, Granville, 224
Sherman, Sandra, 40
Shinagel, Michael, 242n8
Shiner, Larry, 7, 95, 253n3
Silver, Sean, 14–15, 238n66, 251n88
Siskin, Clifford, 247n20
slavery, 54–62, 80, 194–229
Smethurst, Paul, 131, 259n4, 263n71
Smith, Adam, 16, 41, 162–64, 196, 198, 205, 265n17, 269n12
Smith, Bernard, 259n4
Smith, Charles Saumarez, 254n15
Smith, Chloe Wigston, 92
Smith, John, 5
Smith, Pamela H., 10, 13, 235n24, 236n25, 262n51
Smith, Vanessa, 144, 261n42
Snodin, Michael, 259nn101–3
social knowledge, 2, 4, 23–26, 39, 49, 57, 62, 125–26, 129, 148–55, 168–69
Society for the Encouragement of Arts, Manufactures and Commerce (Society of Arts), 95–96, 97, 107, 253nn5–6
Society of Antiquaries, 161
sociology, 2, 23–26

Socrates. *See* Xenophon
Solander, Daniel, 134, 138, 139, 240n94, 265n10
Sollors, Werner, 271n41
Sorensen, Janet, 243n24
Spence, Joseph, 64–66, 74, 247n11, 247n16
Spillers, Hortense, 206, 272n56
spiritual autobiography, 211
Sporing, Herman, 138
Sprat, Thomas, 2–3, 9–11, 17–19, 22, 68, 81, 107
Stafford, Barbara Maria, 151, 263n67
Staley, David, 230
Sterne, Laurence, 45, 259n100
Sudan, Rajani, 5
Swift, Jonathan, 33, 65–66, 86, 90–91, 239n72
sympathy, 214–16

Tasterstal, Robert, 65
Taussig, Michael, 152, 204, 208, 263n68
techne. See knowledge, practical
Tedlock, Barbara, 261n44
theory/practice binary, the, 3, 8, 10, 14, 25, 26, 230–32; Addison, Joseph and, 2, 233n2; and anthropology, 25, 142; Arendt, Hannah and, 228; Aristotle and, 8, 97, 109, 182; Bacon, Francis and, 8, 35, 39, 233n2; Bourdieu, Pierre and, 126, 241n100; Collier, Mary and, 66; Defoe, Daniel and, 35, 39, 41; Duck, Stephen, and, 6, 66; and natural philosophy, 7, 8, 14; and how-to manuals, 122; Hogarth, William, and, 98, 254n20; Johnson, Samuel and, 158, 160, 168; the Royal Society and, 10–12, 231
Thomas, Nicholas, 147, 152, 259n5
Thompson, E. P., 77
Thompson, Helen, 51, 240n82, 244n45
Thomson, James, 70
Thornhill, James, 96–97, 254n15
Tickell, Thomas, 73–74
trades, 164; and artisanal knowledge, 33, 66; artisan-tradesmen, 5, 10, 32–36,

42; and the "consumer revolution," 5, 235n14; bookkeeping, 39–40; craft guilds, 235n14; Defoe's *Complete English Tradesman*, 26, 32–39, 242n19; experiential education, 33–34, 37, 38; "History of Trades" project, 5, 9, 28, 39, 108, 239n77, 258n91, 261n20; and natural philosophy, 35–37; pickpocketing as a trade, 43, 49, 54, 62; and the plain style, 243n22; tradesman vs. artisan, 37

Trumpener, Katie, 192, 265n17, 268n73

Turnbull, David, 262n458

Turnbull, Gordon, 186, 213

Turnbull, Gordon (18th century), 213, 270n13, 272n67

Turner, James Grantham, 85

Uglow, Jenny, 96, 112, 118, 255n21

use, 98–107, 159, 162–69, 170–74, 184

Van Ghent, Dorothy, 45–46

Van-Hagen, Steve, 74, 249n38, 250n60

Vickers, Ilse, 21

Virgil, *Georgics*: georgic care (*cura*), 93; georgic science, 246n1; husbandry and artisanal knowledge, 27; knowledge of the head and hand, 6; learning from things, 94; the speculative artisan 6; translation by Dryden, John, 6

Wall, Cynthia, 45–46, 241n4, 244n37

Wallis, John, 190

Wallis, Samuel, 135, 136, 137

Warton, Joseph, 74

Watt, Ian, 31, 45, 219, 241n5

Wheatley, Phillis, 219

Wheeler, Roxann, 195

Wilkins, John, 9, 160

William the Lion, King of Alba, 176

Williams, Raymond, 248n34, 249n45, 250n51

Wollheim, Richard, 121

Woodward, John, 160

Xenophon, 101–4, 106

Yahav-Brown, Amit, 245n56

Young, Iris Marion, 87, 88

Zitin, Abigail, 98, 238n67, 254n20

The authorized representative in the EU for product safety and compliance is:
Mare Nostrum Group
B.V Doelen 72
4831 GR Breda
The Netherlands

www.ingramcontent.com/pod-product-compliance
Lightning Source LLC
Chambersburg PA
CBHW031800220426
43662CB00007B/478